Geotechnical Centrifuge Technology

Geotechnical Centrifuge Technology

Edited by

R. N. TAYLOR
Geotechnical Engineering Research Centre
City University
London

CRC Press
Taylor & Francis Group
Boca Raton London New York

CRC Press is an imprint of the
Taylor & Francis Group, an **informa** business

A TAYLOR & FRANCIS BOOK

CRC Press
Taylor & Francis Group
6000 Broken Sound Parkway NW, Suite 300
Boca Raton, FL 33487-2742

First issued in paperback 2019

© 1995 by Taylor & Francis Group, LLC
CRC Press is an imprint of Taylor & Francis Group, an Informa business

No claim to original U.S. Government works

Typeset in 10/12 pt Times by AFS Image Setters Ltd, Glasgow

ISBN-13: 978-0-7514-0032-8 (hbk)
ISBN-13: 978-0-367-86385-2 (pbk)

A catalogue record for this book is available from the British Library

Library of Congress Catalog Card Number: 94–72606

Publisher's Note
The publisher has gone to great lengths to ensure the quality of this reprint but points out that some imperfections in the original may be apparent

**Visit the Taylor & Francis Web site at
http://www.taylorandfrancis.com**

**and the CRC Press Web site at
http://www.crcpress.com**

Preface

Centrifuge testing concerns the study of geotechnical events using small scale models subjected to acceleration fields of magnitude many times Earth's gravity. With this technique, self weight stresses and gravity dependent processes are correctly reproduced and observations from small scale models can be related to the full scale prototype situation using well established scaling laws. Centrifuge model tests have proved to be particularly valuable in revealing mechanisms of deformation and collapse and in providing data for validation of numerical analyses. The purpose of this book is to provide both an introduction and a guide to all aspects of centrifuge model testing, from basic considerations of model tests through to applications in geotechnical engineering.

The book comprises nine chapters. The first three chapters concern the background to centrifuge testing, including an historical background, basic scaling relationships and model testing techniques. Examples are then given which illustrate the use of centrifuge model tests to investigate problems often encountered in civil engineering practice. Three chapters deal with aspects of studies on retaining structures, underground excavations and foundations. This is followed by a chapter concerning earthquake and dynamic loading which has become an important element of test programmes for new centrifuge facilities. Finally, there are chapters on environmental geomechanics and transport processes, and cold regions' engineering. This format has been chosen to demonstrate the breadth and powerful nature of centrifuge model testing and its applicability to addressing complex problems in geotechnical engineering.

The book has been written for all geotechnical engineers wishing to learn about centrifuge testing. It is a reference text for new and established centrifuge users from universities, research laboratories and industry, and it will be of particular interest to institutions where centrifuge facilities are either available or planned. I hope the reader will find that this book provides a useful and informative insight into the technology and potential of geotechnical centrifuge model testing.

Acknowledgements

Centrifuge model testing demands considerable practical skills and thanks are due to all the technicians who have given invaluable support to the many projects described in this book. I would also like to express my sincere thanks to all the contributors for their time and effort in making possible the success of this project.

In producing the book, the contributors have drawn on many examples of centrifuge testing. Some of these have been presented at specialist conferences published by Balkema and I am grateful for their permission to reproduce certain figures from those conferences in this book. Publications from which material has been drawn are:

The Application of Centrifuge Modelling to Geotechnical Design. W.H. Craig (editor). Proceedings of a Symposium, Manchester, April 1984. Balkema.
Centrifuge '88. J.-F. Corté (editor). Proceedings of the International Conference on Geotechnical Centrifuge Modelling, Paris, April 1988. Balkema.
Centrifuge '91. H.-Y. Ko (editor). Proceedings of the International Conference Centrifuge '91, Boulder, Colorado, June 1991. Balkema.

These are available from:
A A Balkema
PO Box 1675
NL-3000 BR Rotterdam
The Netherlands

R.N.T.

Foreword

This book gives an account of the principles and practices of geotechnical centrifuge model testing, with nine chapters which review different aspects of the state-of-the-art in 1994. The ten authors of these chapters each worked at an early stage of their research on centrifuges with which I was involved in Manchester or in Cambridge. They have gone on to become authorities on the topics on which they write in this book. There are four groups of chapters: the first three by Craig, Taylor and Phillips give an introduction; the next three by Powrie, Taylor and Kusakabe deal with tests of problems of statics (up to now most tests have been of this type); Steedman and Zeng deal with dynamics; Culligan-Hensley and Savvidou, and Smith consider problems of environmental geotechnics (the newest type of test). This foreword will comment on the background to centrifuge developments. The ISSMFE has a Technical Committee on Centrifuges (TC2) which has held a series of conferences leading to volumes in which the reader of this volume will find further references.

Centrifuge tests are important to geotechnical engineers because there is, in 1994, no universally accepted theory of the strength of soil. We look back to Coulomb, who as a student learned that in addition to friction (Amontons, 1699), soil and rock possessed a property of 'adhesion'. After graduation he made tests which suggested that adhesion in a direct tension test was equal to cohesion in a shear test (Heyman, 1972). Subsequent workers ignored his test data but retained the idea of cohesion in what we call 'Coulomb's equation'. If that equation truly described the strength of soil, then it and the two equations of plane equilibrium of stress would make fields of limiting stress in soil statically determinate, and statical solutions of earth pressure problems could be found by the method of characteristics (Sokolovsky, 1965), without reference to strains.

However, strains do influence earth pressures. One alternative which introduces strains attributes the peak strength of soil in shear to interlocking in the aggregate of soil particles rather than to adhesion between them (Schofield, 1993). In the period from 1954 to 1964, when research workers in soil mechanics in the UK were developing new theories of soil behaviour (Rowe, 1962; Schofield and Wroth, 1968), it was clear that any new theories would have to be tested in application to a wide variety of problems. It was this that led to an interest in centrifuges.

Data from the field are rare and often incomplete. Data from model tests in the laboratory can be misleading because soil self weight in a small model test at the acceleration of normal gravity causes low stress at various points in the model, and the dilation at points in such models is much greater than at homologous points in the field. Sokolovsky's book gave a reference to Pokrovsky's method of centrifuge model testing. Geotechnical centrifuge model tests can give good data, since soil and pore water at homologous points in model and in prototype are at identical stress. In 1964 it was clear that if this method worked, research workers might use centrifuges to test theories of soil behaviour.

A grant was made by the UK Science Research Council and work began on a small centrifuge of 250 mm radius in Cambridge. A package was then made that was taken to Luton and tested on an aerospace centrifuge; Luton were willing to fly the package for 8 hours, after which the package had to be brought back to Cambridge. The tests were concerned with draw down of clay slopes. (Avgherinos and Schofield, 1969), but they also led to insights on the design of centrifuges. This was the subject of a discussion at the Institution of Civil Engineers in London in November 1967, at which I showed four slides (Slides I–IV) which are reproduced here.

Geotechnical engineers are familiar with the dimensionless group which controls slope stability, Slide I, and the idea that the greatest height h of a newly excavated slope will depend on the undrained strength of the soil. The first slide introduced a model of reduced scale at increased acceleration. The second slide, Slide II, showed the dimensionless group that controls consolidation, and introduced the idea that a long process such as staged

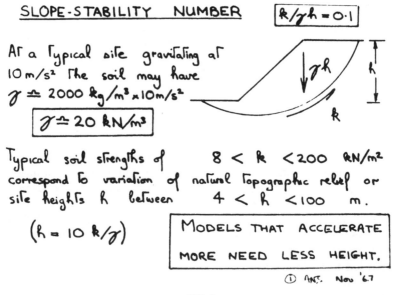

Slide I

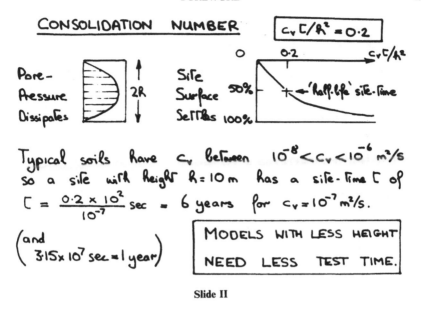

Slide II

construction in the field can be modelled in a short time in a centrifuge. To reduce errors, the model height h should be limited, say to $h = r/10$ (Slide III). Hence, in a stability test of a given slope the strongest soil that can be tested will depend on the velocity, v, that can be achieved in centrifuge flight. The

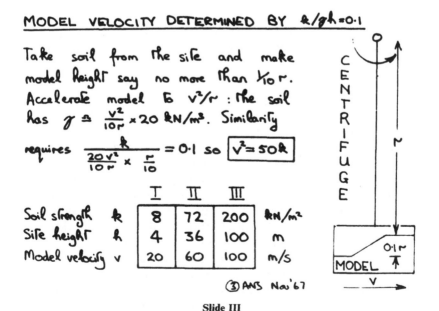

Slide III

MODEL TEST TIME CORRESPONDS TO $C_v t/H^2 = 0.2$ SITE TIME

I Cambridge small centrifuge | $r = 1.4\,m$, $v = 20\,m/s$ |
Model height $2\frac{1}{2}$ cm, $v^2/r = 1600\,m/s^2$ ⎫ Site: 4m bank in soft soil,
Test Time $0.2 \times 10^7 \times 0.025^2$ sec $= \frac{1}{3}$ hour ⎬ Site-Time 1 year

II English Electric (Luton) centrifuge | $r = 2\frac{3}{4}\,m$, $v = 60\,m/s$ |
Model height 27cm, $v^2/r = 1300\,m/s^2$ ⎫ Site: 35m bank in firm soil
Test Time $0.2 \times 10^7 \times 0.27^2$ sec $= 40$ hours ⎬ Site-Time 80 years

III Cambridge-large centrifuge | $r = 3\,m$, $v = 100\,m/s$ |
Model height 30cm, $v^2/r = 3333\,m/s^2$ ⎫ Site: 100m bank in stiff soil
Test Time $0.2 \times 10^7 \times 0.3^2$ sec $= 50$ hours ⎬ Site-Time 700 years

(calculations based on assumed $c_v = 10^{-7}\,m^2/s$ for soil)

(+) AWS Nov '67

Slide IV

speed of sound in air places a limit on the velocity of a beam centrifuge, which can be passed either by operating a beam in a partial vacuum (as now operates in Delft) or with a circular shroud (as now operates for ISMES), or by having a drum centrifuge (as now in Cambridge).

The final slide, Slide IV, reviewed the capabilities of the two centrifuges with which there then had been experience, in Cambridge and Luton, and of a new centrifuge which I envisaged in Cambridge. The slide pointed out that the consolidation half life of a 27 cm clay layer, say 40 hours, was such that we had not been able to reach pore pressure equilibrium in the 8 hour flight that was possible in Luton. At that time it was supposed that clay model preparation in the new centrifuge would take many days, but when the idea of using a downward hydraulic gradient to model self weight of soil was published (Zelikson, 1969), I realised that the same technique could be used to solve the problem of preparation of large centrifuge models.

When I went to Manchester in 1969 I did not know how to build a large drum centrifuge, so I built a centrifuge facility at UMIST with a low cost beam which could be later replaced by a drum. I began drum tests in a small centrifuge borrowed from the Department of Paper Science (English, 1973; Schofield, 1978). Roscoe set the specification for the new Cambridge beam centrifuge to be $r = 4\,m$ and $v = 100\,m/s$, but after his death in 1970 the specification had to be reduced to $v = 70\,m/s$. It was not until 1973 at the ISSMFE Conference in Moscow that I saw Povrovsky's 1968 and 1969 books, in which he reports centrifuges with $r = 3\,m$ radius and $v = 100\,m/s$ as having been used in work on Soviet nuclear weapons.

On my return to Cambridge in 1974 I brought the new centrifuge (the so-called 10 m beam) into operation, and added a swinging platform on which it was possible to place a wide variety of packages and study many different problems (Schofield, 1980, 1981). The former Soviet work had been limited by three factors. They lacked the solid state electronics that is essential in the instruments and computers for data acquisition and signal processing. They taught total stress analysis and they treated soil as viscous rather than elasto-plastic; engineers working in frozen ground may benefit from such teaching but thereafter they consider centrifuge model tests to be crude approximations. Thirdly, the Soviet military value of centrifuge models made this a topic on which discussion was limited. Our success in the past 20 years in the West has depended on modern electronics, effective stress analysis, and (particularly in the conferences on ISSMFE TC2) a free exchange of ideas. Much has been achieved in the past 30 years, as this book shows, and much still lies ahead.

Andrew N. Schofield
Cambridge University Engineering Department

References

Amontons, G. (1699) De la résistance causée dans les machines, tant par les frottemens des parties qui les composent, que par la roideur des Cordes qu'on y employe, et la maniere de calculer l'un et l'autre. *Histoire de l'Académie Royal des Sciences*, **206**, Paris (1702).

Avgherinos, P.J. and Schofield, A.N. (1969) Drawdown failures of centrifuged models. *Proc. 7th Int. Conf. Soil Mech. Found. Eng.*, Mexico, Vol. 2, pp. 497–505.

English, R.J. (1973) Centrifugal Model Testing of Buried Flexible Structures. PhD Thesis, University of Manchester Institute of Science and Technology.

Heyman, J. (1972) *Coulomb's Memoir on Statics: An Essay in the History of Civil Engineering.* Cambridge University Press.

Rowe, P.W. (1962) The stress–dilatancy relation for static equilibrium of an assembly of particles in contact. *Proc. Roy. Soc. Lond.*, **A269**, 500–527.

Schofield, A.N. (1978) Use of centrifuge model testing to assess slope stability. *Can. Geotech. J.*, **15**, 14–31.

Schofield, A.N. (1993) Original Cam-clay. *Keynote Lecture: International Conference on Soft Soil Engineering, Guangzhou*, pp. 40–49. Science Press, Beijing, China. (Also, Technical Report CUED/D-Soils/TR259, Cambridge University Engineering Department.)

Schofield, A.N. (1980) Cambridge geotechnical centrifuge operations. *Geotechnique*, **20**, 227–268.

Schofield, A.N. (1981) Dynamic and earthquake geotechnical centrifuge modelling. *Proc. Int. Conf. on Recent Advances in Geotechnical Earthquake Engineering and Soil Dynamics*, Vol. 3, pp. 1081–1100. University of Missouri-Rolla, Rolla, Missouri.

Schofield, A.N. and Wroth, C.P. (1968) *Critical State Soil Mechanics.* McGraw-Hill, London.

Sokolovsky, V.V. (1965) *Statics of Granular Media* (translated by J.K. Lusher). Pergamon Press, Oxford.

Zelikson, A. (1969) Geotechnical models using the hydraulic gradient similarity method. *Geotechnique*, **19**, 495–508.

Contributors

Dr W.H. Craig   Simon Engineering Laboratories, University of Manchester, Oxford Road, Manchester M13 9PL, UK

Dr P.J. Culligan-Hensley Department of Civil and Environmental Engineering, Massachusetts Institute of Technology, Massachusetts 02139-4307, USA

Professor O. Kusakabe   Department of Civil Engineering, Faculty of Engineering, Hiroshima University, 4-1, Kagamiyama 1-chome, Higashi Hiroshima 724, Japan

Dr R. Phillips   Centre for Cold Ocean Resources Engineering, Memorial University of Newfoundland, St John's, Newfoundland, Canada A1B 3X5

Dr W. Powrie   Department of Civil Engineering, Queen Mary and Westfield College, Mile End Road, London E1 4NS, UK

Dr C. Savvidou   University Engineering Department, University of Cambridge, Trumpington Street, Cambridge CB2 1PZ, UK

Dr C.C. Smith   Department of Civil Engineering, University of Sheffield, Mappin Street, Sheffield S1 3JD, UK

Dr R.S. Steedman   Sir Alexander Gibb and Partners Ltd, Earley House, London Road, Reading RG6 1BL, UK

Dr R.N. Taylor   Geotechnical Engineering Research Centre, City University, Northampton Square, London EC1V 0HB, UK

Dr X. Zeng   Department of Civil Engineering, University of Kentucky, Lexington, Kentucky 40506-0281, USA

Contents

1 Geotechnical centrifuges: past, present and future

W.H. Craig

1.1 Historical perspectives

1.1.1 The germ of an idea and the first experiments (to 1939)

In January 1869 Edouard Phillips presented a paper to the Académie des Sciences in Paris under the title "De l'équilibré des solides élastiques semblables" in which he recognised the limitation of contemporary elastic theory in the analysis of complex structures (Craig, 1989a). Phillips, the son of an English father and a French mother was at the time Chief Engineer of Mines and a teacher at both the École Centrale and the École Polytechnique. His early career working in the railway industry had involved research into the elastic behaviour of steel leaf springs, shock absorbers and beams under both static and dynamic conditions. Faced with intractable analytical problems he recognised the role of models and of model testing.

> Il existe de nombreuses circonstances dans lesquelles les conditions d'équilibre des solides élastiques n'ont pu encore être déduites de la théorie mathématique de l'élasticité, et où elles ne sauraient être obtenues qu'au moyen de méthodes fondées sur des hypothèses plus ou moîs approchées, lesquelles même souvant ne sont pas applicables. Il est donc utile de chercher comment, d'une manière générale, l'expérience peut suppléer à la théorie et fournir à priori, par les résultats de l'observation sur des modèles en petit, les conséquences désirables relatives à des corps de plus grandes dimensions qui peuvent n'être pas encore construits. De là, ce travail qui est basé directement sur la théorie mathématique de l'élasticité.
>
> J'ai traité d'abord, d'une manière générale, et en prenant pour point de départ les équations aux différences partielles fondamentalles, la question de l'équilibre des solides élastiques semblables, supposés homogènes et d'élasticité constante et soumis à des forces extérieures agissant, les unes sur la surface, les autres sur toute la masse, et je me suis proposé la question suivante. Trouver des conditions qui doivent être remplies pour que, dans la déformation, les déplacements élémentaires des points homologues soient parallèles et dans un rapport constant, et qu'il en soit de même pour les forces élastiques, rapportées à l'unité de surface, agissant sur deux éléments superficiels homoloques quelconques, pris dans la masse des corps.

Most importantly he recognised the significance of self-weight body forces in a number of different situations and he developed appropriate scaling relationships from which he developed a recognition of the need for a centrifuge to obtain similarity of stress between models and prototypes when the same materials were used in each.

His initial field of application was to the problems of bridge engineering, which British engineers were at that time pushing forward with bigger and bolder designs. He suggested centrifuging a 1:50 linear scale model of the Britannia tubular bridge over the Menai Straits, to an acceleration of $50\,g$. It was a bold idea—the model itself would have had a length of 8.6 m! He also considered models of a bridge at Conway and of a possible bridge across the English Channel where he considered possible foundation problems.

> Je terminerai cette communication par une remarque. Peut-être les notions précédentes pourraient-elles être appliquées avec fruit dans les études préliminaires d'un project qui, depuis quelques années, a occupé l'attention et qui consisterait à relier la France et l'Angleterre au moyen d'un métallique, reposant sur des piles gigantesques, très-espacées entre elles. Il y aurait là une application, fondée sur des principes rigoureux et analogue aux expériences préliminaires faites avant la construction des ponts tubulaires dont j'ai parlé plus haut, par de célèbres ingénieurs, M.M. Stephenson, Fairbairn et Hodgkinson. Il est permis d'ailleurs de croire qu'on pourrait ainsi éclairer utilement par avance la question de la possibilité de ce projet.

Phillips supplemented his original ideas with another contribution to the Académie, later the same year, in which he considered dynamic effects and showed that in the centrifuge inertial time scaling and linear scaling are in the same ratio between prototype and model. However the idea was apparently doomed to remain in the mind and on paper for some sixty years. No-one in the Victorian era ever applied the centrifuge to the problems of modelling in mechanics.

The first mention of anyone actually undertaking centrifuge modelling appears to be in a paper by Philip Bucky (1931) emanating from Columbia University in the USA relating to the integrity of mine roof structures in rock where small preformed rock structures were subjected to increasing accelerations until they ruptured. Although the work continued at Columbia for some time (Cheney, 1988) there was little or no instrumentation of the models and their significance now is largely historical since there was no mainstream development from this source.

The major early development of geotechnical modelling in the centrifuge occurred in the USSR following independent proposals made by Davidenkov and Pokrovskii in 1932. A number of early papers in the Russian language were available in the West but the first high-profile English-language publication was a presentation by Pokrovskii and Fiodorov made at the First International Conference on Soil Mechanics and Foundation Engineering at Harvard in 1936. Little more was heard of the technique as the century's Second World War developed to be followed by the isolation of the Soviet block behind the so-called 'Iron Curtain'.

1.1.2   The post-war generation (1939–73)

'Re-inventing the wheel' is a somewhat derogatory put down. The rediscovery of the centrifuge must be seen in a different light. It is true that the Columbia machine continued to spin and photo-elastic stress freezing techniques were developed. The US Bureau of Mines investigated rock bolting as an early example of studies of ground–structure interaction to be followed by a number of graduate studies undertaken under the guidance of Professor Clark at the University of Missouri. This work seems to have had limited impact at the time (Cheney, 1988) and activity waned in the face of a developing American preference for mathematical modelling in the age of the digital computer—in effect the stage was empty awaiting new players who would arrive simultaneously from two cultures.

In Japan Professor Mikasa at Osaka City University sought experimental validation of his theory for the consolidation of soft clay deposits in which soil self-weight plays a dominant role (see Kimura, 1988). He appreciated the importance of the self-weight and with appropriate consideration of similarity theory used first a commercially available centrifuge and later one of his own design to support his theoretical work. As his confidence and understanding increased Mikasa added experimental studies on bearing capacity and slope stability to his experience before 1973, developing considerable mechanical and instrumentation expertise. He considered earthquake loading on a slope by rotating his model slope within the centrifuge reference frame using an ingenious mechanism for moving the two main arms of the centrifuge rotor in opposing directions; this was the equivalent of superimposing a one-way static horizontal acceleration on top of a fixed vertical gravitational acceleration as in the usual numerical and analytical approaches at the time.

In England Dr. A.N. Schofield had become aware of the early work in the USSR while translating Russian books on the mechanics of soil. Realising the potential to expand the high-quality, small scale/low stress, model work on structures being undertaken in the 1960s at Cambridge University, he initiated a new programme firstly with a frame mounted on an existing controlled speed turbine in his own laboratories and later using a larger centrifuge designed for environmental testing within the aerospace and defence industries. His early work centred on problems of slope stability in clay soils and necessarily involved considerations of consolidation, thus paralleling the Japanese work in several respects.

In 1969, the International Society for Soil Mechanics and Foundation Engineering (ISSMFE) held its seventh conference in Mexico. For the first time since 1936 there were papers related to centrifuge work—this time from England (Avgherinos and Schofield), Japan (Mikasa *et al.*) and the USSR (Ter-Stepanian and Goldstein) and all devoted to slope stability studies. This is not surprising. While it has to be recognised that the process of movement of pore fluid under the action of gravity-induced hydraulic gradients was crucial in leading the various researchers towards the centrifuge technique, the problem of slope stability was almost inevitably among the initial applications of the new technique. In structures involving slopes, soil self-weight, and possibly self-weight-induced seepage, will be the dominant force causing instability and this can be modelled only in a centrifuge if prototype stress levels are to be preserved. All other geotechnical structures are mechanically more complex, involving elements such as retaining walls, piles, anchors and/or external forms of loading other than soil body forces. At one level at least the slope appears the obvious first target for the centrifuge models to make an impact.

The next ISSMFE conference was held in Moscow in 1973 and it was on this occasion that the extent of Soviet expertise in the field became apparent. Visitors from outside the USSR were invited to meetings and saw for the first time the centrifuges being used there. Professor Pokrovskii was present at one of these meetings and presented copies of two of his monographs to Professor Schofield; these were translated into English (Pokrovskii and Fiodorov, 1975) in due course and although not formally published enjoyed quite wide circulation. It became clear to the new generation of centrifuge enthusiasts that the Russian work in this field had been continuous since the 1930s and a considerable body of experience and expertise had been built up. Much, though by no means all, of the effort had been put into the modelling of such processes as the effects of explosions underground—extent of cratering and ground transmission of vibrations. The obvious military application of this work dictated that it should remain hidden from those beyond the 'Iron Curtain' though it is clear that other work of a non-military nature was also withheld from outsiders in this as in other areas of technology outside geotechnical engineering.

1.1.3   Missionary zeal (1973–85)

Following the Moscow conference there was a rapid advance in knowledge of the technique of centrifuge modelling and also in development of the modelling techniques with associated instrumentation. It is apparent that Russian advances had been made in the absence of the development in numerical computational capability seen elsewhere, but also that their centrifuge technology had been limited by a parallel absence in model instrumentation and data-retrieval techniques. The rapid developments in the fields of instrumentation and computing in other countries led to a period in which both centrifuge modelling and numerical analysis of geotechnical problems developed in parallel. The nature of geotechnical materials with their complex and as yet poorly understood constitutive relationships precluded the rejection of physical experimentation which many numerical hegemonists have repeatedly predicted. While there was huge increase in the number and power of digital computers used in the geotechnical field in the period to 1985 there was also a very significant growth in centrifuge activity by those who recognised the continuing limitations of number only solutions to problems. There was also a conscious effort on behalf of those involved from the start to set out their skills and sell the centrifuge techniques to those who would listen to their message.

The ISSMFE president took the very positive step in 1981 of setting up an international technical committee to promote activity in this field. This committee was run under the auspices of the British Geotechnical Society in the period to 1985 and has subsequently been managed from France to 1989 and the USA to 1994. In the first period a committee of 17 members corresponded widely and arranged three international symposia in the period April–July 1984, held at Tokyo, Manchester and Davis (California). In this one brief period, three substantial published collections totalling some 70 different contributions became available from a total of 20 different groups operating geotechnical centrifuges in 10 different countries. Regrettably, neither of the Russian members of the technical committee was able to attend any of the meetings, nor were they able to contribute in written form.

As a final contribution to their missionary efforts, the technical committee organised a speciality session at the International Conference in San Francisco in 1985 which led to the later publication of a state-of-the-art volume entitled *Centrifuges in Soil Mechanics* which included a collected bibliography of around 400 publications in this fast developing field (Craig *et al.*, 1988).

1.1.4 Recognition and the years of production

After 1985 the existence of centrifuge modelling capabilities in a good number of countries was widely recognised. Since then there has been an increased acceptance of the technique by practising engineers and continued increases in the size and number of machines available as well as a widening range of operators. In the UK, all centrifuges continue to operate within the university sector but this is not always the case elsewhere. Non-university research laboratories and comparable organisations have funded nationally operated machines in a number of countries, e.g. France, Netherlands. In the USA there are now a number of machines of varying size in the university sector, some of these receiving very considerable central funding and enjoying 'national' status while the military operate some others. The US Army is currently building what will probably be the largest geotechnical centrifuge in the world with a specification and potential capability which open up new avenues for the future in both military and civilian applications. In Japan there has been continuous development since the initial experiments of Mikasa and the number and variety of machines continue to expand, encompassing the research community but perhaps more significantly the commercial sector. It is only in Japan that the practising engineering community has fully embraced the potential of this powerful experimental approach within the integrated range of available tools.

Since 1985 the Technical Committee has organised further International speciality conferences in Paris (1988), Boulder, Colorado (1991) and Singapore (1994). These have each spawned conference volumes containing around 80 papers in the specific field of centrifuge modelling. The meetings have attracted principally those already involved in this particular market sector and initial inspection of the volumes (Corté, 1988; Ko and McLean, 1991) reveals much concentration on experimental techniques and hardware. Closer study reveals also the rapidly widening range of application of the modelling techniques and this range is demonstrated elsewhere in the present volume.

The speciality conference will always contain a strong element of preaching to the converted and there has been an overt recognition of the need to be less introspective and to demonstrate to the wide engineering profession what capabilities are on offer and available. An increasing proportion of the output from the centrifuge community is now being disseminated in the more general literature where its acceptance demonstrates the status now achieved by this technology.

1.2   Developments in hardware

1.2.1   Machine configurations

Without exception, all the machines currently operational rotate in a horizontal plane with a drive system acting on the axis of rotation. There have in the past been other configurations under consideration including small centrifuges rotating in a vertical plane and rail- or sled-mounted model containers powered around the walls of a circular enclosure. The first of these is impractical at any large size because of the difficulties in coping with the downward gravitational acceleration on a model passing through the top of the circles of rotation at low speed and the cyclical $\pm$ 1g component superimposed on the steady centrifuge acceleration at any speed. The rail- or sled-mounted model container is superficially attractive but leads to serious problems associated with heavy moving point loads on the outer peripheral structure and with control and instrumentation within the model itself which require either telemetric systems or a nominal structural link back to rotating joints and slip-rings on the central axis.

Within the existing family of horizontal plane, axially driven centrifuges there are a number of identifiable sub-groups, namely: (i) rotating arm— symmetric/asymmetric designs, and hinged/rigid/hybrid model containers; and (ii) rotating drum. The majority of machines are of the rotating-arm type. Many of the older designs are based on the principle of using a symmetric, balanced beam with the possibility of simultaneously using two models of comparable size and mass, one at each end of the arm. While superficially attractive, this possibility has been rarely exercised in practice because of the desire to concentrate efforts in specimen preparation and instrumentation on a single model and the inevitable compromises and conflicts involved in dual model testing. Machines of this structural form virtually always operate with inert counterweights balancing the active model assemblies. Examples in the UK are machines in Cambridge and Manchester and in Japan at Osaka—all were designed around 1970.

In contrast to this concept, more recent designs have recognised the economic advantages of gaining a machine requiring less power for given model size, radius of rotation and level of acceleration by using an asymmetric arm with model container at a long radius balanced by a more massive counterweight at a smaller radius. As with the balanced arm designs, it is possible nowadays to adjust the balance under static conditions before starting the centrifuge, to check for minor levels of imbalance during machine rotation and to correct for unacceptable conditions by moving solid masses mechanically or by pumping liquids between balance tanks. Such systems are included to protect the structural integrity of the machines and to limit wear on bearings.

With either of the above design concepts it is possible to use an end-of-arm arrangement which is rigid, hinged or a hybrid. The rigid arm structure is simplest and probably cheapest in every case but suffers from a similar, if less acute, disadvantage to that of adopting the vertical plane of rotation. When a model is mounted on a rigid arm machine, the gravitational (vertical) acceleration is orthogonal to the centrifuge acceleration (radial/horizontal). Fluids and cohesionless soils will thus not remain in their required positions while the centrifuge is stationary unless some artificial restraint is provided. As centrifuge rotational speed is increased the equipotential surfaces for liquids freely enclosed in a model box tend to the paraboloid of rotation, which approximates to a cylinder at high acceleration levels. Models designed to be tested at a particular linear scale ($1:N$) under an acceleration which is fixed at Ng can be built with 'level' surfaces inclined to the axis of rotation at a slope of $1:N$ and curved in the plane of rotation. This is particularly necessary in large three-dimensional models and indeed is usually done. With two-dimensional, plane models there are two options, namely to test with the model plane in the plane of rotation or orthogonal to it. The distortions from ideal geometric similarity with a prototype are minimised by using the latter arrangement and this is to be preferred in most instances; prototype planar surfaces remain planar in the model and there is no difficulty with strain distortion. The disadvantage with this approach lies in the need for a larger frontal area to be presented to the direction of rotation than is the case where a narrow two-dimensional model lies in the horizontal plane. In this alternative arrangement the model horizontal or 'level' planes are curved and vertical planes are radial with curvature and angular separation depending on the ratio of model length to radius of rotation. Taking as an example a model 1 m wide rotating at a nominal radius of 5 m, the angle between extreme radial lines is 11.4° and the level surface departs from a straight line by 25 mm. If model 'verticals' are indeed made radial then a 10 mm uniform settlement of the surface induces a tensile strain of 0.2% which may result in tension cracking. In practice there appear to be no instances where model verticals have been made strictly radial—all containers used, even in this orientation have been rectilinear. Some compromises have to be made with either orientation of the plane model, but under static accelerations the balance of advantages favours having the model plane parallel to the axis of rotation. When there is a desire to model dynamic behaviour, further considerations come into play but the same overall conclusion will remain.

Much more common than the rigid rotating arm, is the family of designs, symmetric or asymmetric, in which the model container is mounted on a swinging platform attached to one end of the rotor by a hinge or hinges. Such a platform will ideally lie so that the base of the platform will always be normal to the resultant of the gravitational and centrifugal accelerations, i.e. horizontal under static conditions but close to the vertical cylinder of rotation at high centrifugal accelerations. Such designs increase the complexity and cost of the rotor itself but have the clear advantage over the rigid form that cohesionless soils and fluids can be accommodated at all times without artificial restraints. In the vast majority of current machines, the model support platform has essentially equal orthogonal dimensions yielding a ready accommodation of square and circular model containers as well as of two-dimensional containers with orientations either parallel to, or normal to, the axis of rotation.

There are other possibilities for hybrid variations based on essentially rigid arm machines. If the model platform on a rigid arm is large enough it is possible to mount a smaller swinging model container of the rigid base as has been done in Manchester (Craig and Wright, 1981). At Cambridge the centrifuge arm itself is rigid but an articulated model container has been added (Schofield, 1980) which swings freely under static conditions and at low g but is seated back onto the rigid arm end face plate before high g conditions are reached—thus combining the advantages of both designs and allowing a novel approach to simulating dynamic inputs (see section 1.2.3).

The alternative machine configuration to the rotating arm is the rotating drum in which a soil specimen is placed along the full periphery of a cylinder rotating about its axis. Such machines have always been a minority and the possible applications have yet to be fully explored. The problems described above associated with two-dimensional models mounted in the plane of rotation clearly exist with the drum centrifuge format, but there are compensating advantages associated with the great increase in one model dimension which can be considered virtually infinite in some respects.

Early models with drum centrifuges were concentrated on slope stability problems (e.g. Cheney and Oskoorouchi, 1982) and small machines with radii less than 1 m have rotated about a horizontal axis. In such instances, soil beds can be prepared by consolidating material added as a slurry or as a dry cohesionless material to the already rotating drum container. Once the rotating bed has been formed around the periphery, excavation of soil can be simulated by removing material using a stationary tool introduced from within the drum, as in the use of a lathe. Alternatively, continuous embankment loading can be simulated by the steady introduction of further cohesionless material from a stationary point.

More sophisticated machinery (e.g. Sekiguchi and Phillips, 1991) permits a much more versatile use of the drum centrifuge. When an axial assembly within the drum rotates at the same speed as the drum then actions can be initiated which occur at specified locations around the periphery rather than continuously around it. This may not appear particularly advantageous since this is more easily done on a rotating-arm structure, but if more than one action point can be engineered, or if a single point can be displaced in stages relative to the drum whilst the centrifuge is rotating, then multiple tests can be carried out at discrete locations on a single large soil bed. If continuous relative motion can be achieved between axial assembly and periphery, then such actions as ploughing, dredging and movement of continuous ice sheets past fixed structures can be simulated. Considerable future development can be expected with this type of machinery by the turn of the century.

The mention above of ice sheets introduces a further area of centrifuge hardware development—that of environmental simulation. There are as yet few instances of the use of fully enclosed environmental chambers allowing for the simulation of controlled high- or low-temperature regimes though a number of experiments have been reported in which thermal gradients have been generated by remotely controlled heating elements (e.g. Savvidou, 1988) and ice layers have been formed by the controlled supply of cold gases to model surfaces (Vinson, 1983; Lovell and Schofield, 1986). Similarly effects akin to precipitation induced ingress of water to soil surfaces have been simulated (Kimura et al., 1991; Craig et al., 1991)—if not yet very precisely.

One major centrifuge has been designed to operate within an enclosure which can be partially evacuated in order to reduce drive power requirements (Nelissen, 1991). In this scenario two possibilities exist, namely maintaining normal atmospheric pressure within a model container or allowing the soil fluid pressures to be linked to the reduced external atmospheric pressure. In the latter case absolute pressure levels will be reduced and due consideration has to be given to the effects this may have where pore suctions are operative (relative to gauge pressure) and in all cases of partially saturated soils.

1.2.2 Consideration of linear scale

1.2.2.1 Small structures in civil engineering. Much of the development in geotechnical centrifuge modelling has been associated with problems in the civil engineering or construction industries where structural elements of limited size are involved. Many piles are of the order of 250–1000 mm in width or diameter and typical concrete elements such as retaining walls have comparable dimensions. Similarly pipes, culverts, etc. have limited critical dimensions and for realistic modelling it is often appropriate to select a linear scale factor in the range 20–25. An increase in scale factor of 2 leads to a reduction in overall model mass by a factor of 8, i.e. a much smaller and lighter model requiring a smaller but higher speed centrifuge. As in all modelling situations, an optimisation is required balancing the costs and benefits of larger and smaller models. In this class of structure larger models and limited scaling ratios are commonly adopted.

1.2.2.2 Modelling of construction.   Much of the practice of geotechnical engineering depends upon the manner of the construction process—pile driving, compaction, backfilling of trenches, placement of fill, pattern of excavation. Many such processes have already been simulated aboard centrifuges to varying degrees of realism. Various pile-installation techniques have been modelled (compare Nunez and Randolph, 1984; Oldham, 1984; Sabagh, 1984) and there have been studies comparing the performance of piles installed in soil beds in the laboratory and subsequently loaded under centrifuge accelerations with similar piles installed, by pushing or dynamic driving, in a high acceleration field (Craig, 1985). The results clearly indicate that in certain circumstances there is a need to simulate the process of construction and not merely the final overall geometry if realistic prototype modelling is to be achieved. As centrifuge experiments become more sophisticated and ambitious it is apparent that there will be further developments in this area.

To date, a number of techniques have been used to simulate the stress changes involved in excavation—removal of mechanical support (Craig and Yildirim, 1976), drainage of fluids, lifting of bags or blocks of soil (Azevedo and Ko, 1988). No-one has yet simulated the methods of excavation normally used in the field. Many experimenters have simulated the placement of cohesionless fills by raining dry sand from storage hoppers mounted above a rotating model (e.g. Beasley and James, 1976). This is a good first step but there remains much development before model fills experience the stress regimes associated with roller or vibratory compaction in the field. Cohesive materials present even more intractable difficulties.

The technique of pouring or raining sand from model hoppers is a direct analogue of the processes involved in hoppers and storage silos and such field processes as pressure surging in grain or sand storage containers can and have been modelled (Craig and Wright, 1981; Nielsen, 1984). Grain size effects become very significant if flow orifices are small and minor changes in moisture content become very significant—but no more so than in the field.

In all these process-dependent models, there is again a clear trend to the use of relatively large models, i.e. to limited scale ratios. Small is not necessarily beautiful and the larger size, lower acceleration centrifuges have advantages.

1.2.2.3 Larger sites and regional problems.   When global site assessment is not likely to be governed by the performance of small components, then the use of higher scale ratios has obvious attractions. It is significant to note that most centrifuges built in the last decade have maximum accelerations of 200 g or more. The machine sizes vary widely but there is much consistency in the selection of acceleration parameter. A review of published work (Craig, 1991) suggests that the machines do not often run at these maximum accelerations—the arguments above, favouring limited accelerations, are widely accepted. However, the original machine specifications are fully justified by the, possibly rare, occasions when it is appropriate to combine large models and high accelerations to simulate sites of a maximum possible size in which global structural behaviour is dominated by continuum rather than elementary or particulate mechanisms. Obvious examples are in geotectonic studies, the performance of large structures under static or dynamic loadings and the flow of ice sheets around structures.

It is common practice to invoke two justifications for centrifuge modelling; firstly the benefits of stress equality with the prototype combined with the convenience of reduced model size and secondly the benefits of reduced time in small-scale models—see chapter 2. There are problems associated with time scaling of different phenomena and in particular in meeting the twin requirements of simultaneously satisfying the similarity laws for inertial and drainage effects. Over a limited range of scale ratios it is possible to meet the resulting dichotomy by the use of replacement pore fluids other than water, but there are limits to this technique which in any event is only applicable to clean cohesionless soils.

1.2.3 Loadings other than gravity

While gravity exerts a dominant effect in all geotechnical situations there are often other forces and loadings which are significant. Vehicle braking forces may be imposed on bridge piers and abutments, ice or fluid flows cause lateral loadings on piles—such loadings can often be simulated by mechanical devices aboard a centrifuge. Other loadings such as blast, impact and seismic loading require different approaches. Wave loading of offshore structures was the basis of a great deal of centrifuge work in the 1970s and 1980s. The whole gamut of offshore structure platforms has been simulated—gravity platforms, piled structures, jack-ups, components of tension leg platforms. In virtually every case, wave loads were simulated mechanically and model structures were subjected to various static and cyclic loading or displacement patterns produced by pneumatic rams, electric motors or hydraulic actuators. The Manchester centrifuge group utilised a large payload capacity to pioneer the use of a range of heavy servo-hydraulic actuators under computer control which provided a versatile and very powerful loading capability which was able to meet a wide range of different requirements on models of large size, with minimum rejigging between projects (Craig and Rowe, 1981). This technology would later be widely used elsewhere in the simulation of earthquakes. Other groups used low power, but lighter systems combined with smaller models.

A number of experimenters have detonated explosive charges within models in order to simulate a wide range of explosive events. As indicated above, developments in the Soviet Union were the first in this area of activity exploring both civilian and military objectives (Pokrovskii and Fiodorov, 1975). Studies there and elsewhere have fallen into two main categories—those associated with the large-scale effects of explosions, e.g. assessment of crater volume and optimisation of combinations of charge depth and size, and those associated with blast-induced effects on structures. The first type of experiment has generally been carried out at quite high-scale factors and large explosive yields in the prototype have been simulated (Schmidt, 1988). Such experiments have been mechanically simple, involving the necessary safety requirements and a controlled detonation circuit but minimal instrumentation. More sophisticated blast-wave propagation studies and assessment of structural performance require more complex instrumentation on carefully detailed models and have generally been carried out at lower acceleration levels (e.g. Townsend et al., 1988)—as ever, where detail matters the scale factors have been limited.

Blast models and those involving single impacts may involve the determination of wave refraction and reflection effects in certain instances but generally the time of interest within the model is that associated with the first passage of a shock wave—boundary effects are, as in all models, of some concern but are not generally critical.

Much more complex than blast and impact models are those involving simulation of repeated dynamic loadings and seismic loadings which have become increasingly common in the last decade. In such models the mechanics of generating the required dynamic inputs to the model are complex and there is a need to consider boundary effects and boundary-wave reflections in most cases. These problems are detailed in other chapters of this book but it is pertinent to introduce them here since basic model principles, having an impact on hardware design back to basic centrifuge configurations, are involved.

Considering container boundary effects first, there remain two approaches to the problem of simulating the performance of something as apparently simple as a level bed of uniformly deposited, saturated cohesionless soil subjected to base shaking. Most simply, a perfectly rigid container can be shaken by whatever mechanism is conceived. In this case the side walls of the container move with the base and the resulting shear strains within the container will not be uniform at any horizon. Displacements may be small but the concept of the model as a shear beam is flawed. When more complex model situations are involved, possibly involving non-level horizons and structural inclusions then the wave transmission is not even idealised as one-dimensional and there is a widely recognised need to damp-out wave reflections from model boundaries. This has been achieved with an apparently high degree of success by lining model container walls with vibration suppressing materials (e.g. Cheney et al., 1988; Campbell et al., 1991)—a similar approach has been used in blast and impact work where container bases are also lined.

An alternative approach has been to adopt the principle of a perfectly flexible model container which is able to distort laterally in response to base input accelerations. Such containers have become known as 'stacked-ring' or articulated models and are built up from a series of rigid support structures, of small depth, each having high inherent stiffness but assembled in such a way as to be able to move laterally without interference (e.g. Law et al., 1991). This appears to offer considerable advantages in allowing freedom of a model to move in ideal ways, with a minimum of constraint which is absent in the field. Such containers can generally be smaller than rigid ones for a comparable model performance, since there is no need to allow for a sacrificial dead zone close to the side walls as is necessary in the latter. However, there are obvious penalties in direct costs of model construction and assembly and the articulated models are much more difficult to use and require much maintenance and considerable skill and ingenuity. Several groups working in this field have used both approaches and there is, as yet, no consensus that one is necessarily better than the other. Pragmatism and idealism are here in conflict and each model scenario requires individual consideration.

The simulation of seismic or other comparable dynamic inputs to models has generated much effort in the last decade or so. Repeated cyclic vibrations have been generated by triggering natural vibrations of models displaced from equilibrium while supported using flexible suspensions (Morris, 1983). This is simple and cheap. The Cambridge centrifuge group have used the so-called 'bumpy road' device in which a roller on the rotating arm of the centrifuge has been brought into contact with the rigid wall of the rotor pit (Kutter, 1982). The wall has been profiled to transmit a predetermined (sinusoidal) lateral motion into a container mounted on the otherwise rigid rotating arm. The dynamic forces transmitted into the rotor structure are quite modest but generally unquantified. The basal accelerations input to either an articulated or rigid model container can be quantified and have been found to be quite repeatable. The technique is restricted to acceleration–time inputs of limited form and to accelerations only in the plane of rotation—thus the comments above about model orientation apply once more. Notwithstanding these limitations the technique had been famously successful and has generated a great deal of insight into the performance of simple soil beds and of more complex soil–structure configurations under repeated inertial accelerations.

True seismic simulation requires an ability to generate inputs which are closer to reality, involving multi-frequency components possibly in more than one direction and the techniques of spectral analysis of both inputs and outputs. Most groups moving into this field now adopt servo-hydraulic power systems to generate digitally controlled input accelerations. The basic technology has been used for many years in the simulation of low frequency cyclic inputs associated with modelling water wave loading on offshore structures. The need with seismic modelling is for short bursts of high-power input and has to be met by the inclusion of hydraulic accumulators on the centrifuge close to the model. The components of such systems are of necessity quite large and massive and while high-quality reproductions of field-recorded earthquake records have been reproduced on a single axis in quite small models there is not yet any experience of generating three-dimensional or even two-dimensional inputs. The mechanical and control engineering aspects of this work are difficult but further developments are to be expected. One aim must be to increase the proportion of soil mass within an overall model assembly which is currently typically around 25%. While acceleration records can be reproduced on a single axis there are inevitably cross-accelerations generated on others, but these have rarely been measured or acknowledged—this problem will have to be addressed on multi-axis shaking systems. Notwithstanding these problems, increasing expertise and ingenuity can be expected to deliver new insights and understanding.

1.3   Where will it end?

The range of centrifuge machinery, ancillary equipment and instrumentation continues to expand together with the cumulative confidence and versatility of the experimenters. Over the years several authors have categorised centrifuge usage in various ways. The author has previously highlighted four areas of activity where the centrifuge can make significant contributions (Craig, 1988), these being:

1. Teaching.
2. Development of an understanding of mechanics of geotechnical materials and structural forms.
3. Development of new methods of analysis (or modification of existing methods) for idealised, but realistic, geotechnical structures by providing quantitative indications of the effects of parametric variations once the overall mechanics are established.
4. Modelling of site-specific situations to assist in design studies and project appraisal.

A review of present practice indicates that all of these are currently active areas.

    The use of the centrifuge as a tool for instruction of students grappling with the complexities of soil behaviour and soil–structure interactions has become increasingly recognised in a number of countries. It is to be hoped and expected that small cheap machinery will be developed and widely used in this role in education institutions which have no record of research or professional practice in the use of centrifuges (Craig, 1989b). Geotechnical students have for a generation been confronted by the standard consolidation and shear-strength experiments in laboratory classes. How much more valuable to them is a well-designed demonstration using centrifuge models, with or without the embellishment of instrumentation at an appropriate level?

The development of an understanding of mechanics in normal forms of construction and under unusual loadings is of the essence in much centrifuge work. One need only look back at the insights given in the fields of offshore foundations and seismic liquefaction to see the immediate impact centrifuge modelling has had on engineering practice. It is difficult to be specific when looking forward beyond the horizon but major further advances can be expected in seismic studies provided a generation of multi-axis shaking platforms is developed. The development of a large centrifuge in Canada specifically for cold regions' engineering work suggests there will be rapid advances in the fields of ice mechanics and frozen ground. Construction of an even bigger machine in the mid-1990s at the Waterways Experiment Station in the USA will permit modelling of a wide range of phenomena involving simulated depths greater than have generally been available to date—this is another exciting prospect. Multiple developments in the commercial sector in Japan suggest that novel construction techniques will be subject to rapid model study appraisal at the development stage. Much innovation in the practice of ground engineering has been 'contractor driven' and the inclusion of model work as a contributor to this innovation process is to be welcomed.

The use of the model test in developing new methods of analysis has a wider scope than the 'validation of numerical methods' or 'checking of computer codes' which is sometimes cited as a role. Experience with the multi-million dollar VELACS project (VErification of Liquefaction Analysis by Centrifuge Studies) in the period 1989–93 has shown how such a role can be coordinated. It is apparent that in relatively complex situations the state-of-the-art calculations cannot simply be proved or validated by checking against model data— there needs to be an iterative process whereby the physical model data, which themselves may need to be checked for repeatability or consistency between different sources, contribute to the ongoing development of constitutive modelling laws which are themselves merely components to be built in the codings which have to be robust and adaptable. This sort of work will continue for some time yet and just as computational expertise develops so will that in physical modelling. Though the inflation of costing may dissuade some from undertaking the latter, the world of geotechnical engineering cannot yet afford to abandon the tangible model for the superficial attraction of the intangible which may appear to be cheaper.

Consideration of the possibilities of modelling site-specific situations tends to provoke impassioned debate amongst centrifuge aficionados. By definition, a model will not represent all features of a prototype under consideration during a design study or under construction—this is true of a physical model just as for a numerical model. Nevertheless, it is apparent that provided essential features (including geotechnical materials) are identified and included in a model to a satisfactory degree of similarity then physical model data can, and should, usefully contribute to design which is an interactive process involving the synthesis of many inputs. Model studies can provide indications of 'most probable' performance, can contribute to determining 'worst acceptable' scenarios and to the assessment of possible responses to varying performance when the 'observational method' is used in practice.

There seems likely to be a continued widespread use of centrifuge modelling into the next century in many roles. As the range of centrifuges continues to expand, basic machinery is being developed to cover an ever-widening envelope of possible model situations. As the pool of experienced experimenters widens and develops and as the engineering of ancillary equipment becomes more sophisticated, the versatility of the technique will expand. Some caution must be exercised. As the corporate confidence of the pool of modellers increases so does their ambition. To date there have been a number of serious, but non-catastrophic, incidents in centrifuge operations. As machines became bigger and more powerful and experiments possibly become more aggressive (blast, vibration, etc.) the potential for a major accident increases and the centrifuge operators must be ever vigilant in trying to avoid disaster. In the continuing expansion of confidence, they will need to remain vigilant that their results are only used where principles of similarity have been rigorously proved and the limitations of the data produced are understood by the end users.

References

Avgherinos, P.J. and Schofield, A.N. (1969) Drawdown failures of centrifuged models. *Proc. 7th Int. Conf. Soil Mech. Found. Eng.*, Vol. 2, pp. 497–505.
Azevedo, R. and Ko, H.Y. (1988) In-flight centrifuge excavation tests in sand. *Proc. Conf. Centrifuge '88*, pp. 119–124. Balkema, Rotterdam.
Beasley, D.H. and James, R.G. (1976) Use of a hopper to simulate embankment construction in a centrifugal model. *Géotechnique*, **26**, 220–226.
Bucky, P.B. (1931) Use of models for the study of mining problems. *Am. Inst. Min. Met. Eng.*, Tech. Pub. 425, 28 pp.
Campbell, D.J., Cheney, J.A. and Kutter, B.L. (1991) Boundary effects in dynamic centrifuge model tests. *Proc. Conf. Centrifuge '91*, pp. 441–448, Balkema, Rotterdam.
Cheney, J.A. (1988) American literature on geomechanical centrifuge modelling 1931–1984. In *Centrifuges in Soil Mechanics* (eds W.H. Craig, R.G. James and A.N. Schofield), pp. 77–80. Balkema, Rotterdam.
Cheney, J.A. and Oskoorouchi, A.M. (1982) Physical modelling of clay slopes in the drum centrifuge. *Transport. Res. Rec.*, 872, pp. 1–7. Washington D.C.

Cheney, J.A., Hor, O.Y.Z., Brown, R.K. and Dhat, N.R. (1988) Foundation vibration in centrifuge models. *Proc. Conf. Centrifuge '88*, pp. 481–486. Balkema, Rotterdam.

Corté, J.F. (ed.) (1988) *Proc. Conf. Centrifuge '88*. Balkema, Rotterdam.

Craig, W.H. (1985) Modelling pile installation in centrifuge experiments. *Proc. 11th Int. Conf. Soil Mech. Found. Eng.*, Vol. 2, pp. 1101–1104.

Craig, W.H. (1988) On the uses of a centrifuge. *Proc. Conf. Centrifuge '88*, pp. 1–6. Balkema, Rotterdam.

Craig, W.H. (1989a) Edouard Phillips (1821–1889) and the idea of centrifuge modelling. *Géotechnique*, **39**, 697–700.

Craig, W.H. (1989b) The use of a centrifuge in geotechnical engineering education. *Geotech. Test. J.*, **12**, 288–291.

Craig, W.H. (1991) The future of geotechnical centrifuges. *Proc. ASCE Geotechnical Congress*, Vol. 2, pp. 815–826.

Craig, W.H. and Rowe, P.W. (1981) Operation of geotechnical centrifuge from 1970 to 1979. *Geotech. Test. J.*, **4**, pp. 19–25.

Craig, W.H. and Wright, A.C.S. (1981) Centrifugal modelling in flow prediction studies for granular materials. *Particle Technology*, Inst. Chemical Engineers, pp. D4, U1–14.

Craig, W.H. and Yildirim, S. (1976) Modelling excavations and excavation processes. *Proc. 6th Eur. Conf. Soil Mech. Found. Eng.*, Vol. 1, pp. 33–36.

Craig, W.H., James, R.G. and Schofield, A.N. (eds) (1988) *Centrifuges in Soil Mechanics*. Balkema, Rotterdam.

Craig, W.H., Bujang, B.K.H. and Merrifield, C.M. (1991) Simulation of climatic conditions in centrifuge models tests. *Geotech. Test. J.*, **14**, pp. 406–412.

Kimura, T. (1988) Centrifuge research activities in Japan. In *Centrifuges in Soil Mechanics* (eds. W.H. Craig, R.G. James and A.N. Schofield), pp. 19–28. Balkema, Rotterdam.

Kimura, T., Takemura, J., Suemasa, N. and Hiro-oka, A. (1991) Failure of fills due to rainfall. *Proc. Conf. Centrifuge '91*, pp. 509–516. Balkema, Rotterdam.

Ko, H.Y. and McLean, F.G. (eds) (1991) *Proc. Conf. Centrifuge '91*. Balkema, Rotterdam.

Kutter, B.L. (1982) Deformation of centrifuge models of clay embankments due to 'bumpy road' earthquakes. *Proc. Conf. on Soil Dynamics and Earthquake Engineering*, Vol. 1, pp. 331–350. Balkema, Rotterdam.

Law, H., Ko, H.Y., Sture, S. and Pok, R. (1991) Development and performance of a laminar container for earthquake liquefaction studies. *Proc. Conf. Centrifuge '91*, pp. 369–376. Balkema, Rotterdam.

Lovell, M.S. and Schofield, A.N. (1986) Centrifugal modelling of sea ice. *Proc. 1st Int. Conf. on Ice Technology*, pp. 105–113. Springer-Verlag, Berlin.

Mikasa, M., Takada, N. and Yamada, K. (1969) Centrifugal model test of a rockfill dam. *Proc. 7th Int. Conf. Soil Mech. Found. Eng.*, Vol. 2, pp. 325–333.

Morris, D.V. (1983) An apparatus for investigating earthquake-induced liquefaction experimentally. *Canad. Geotech. J.*, **20**, pp. 840–845.

Nelissen, H.A.M. (1991) The Delft geotechnical centrifuge. *Proc. Conf. Centrifuge '91*, pp. 35–42. Balkema, Rotterdam.

Nielsen, J. (1984) Centrifuge testing as a tool in silo research. *Proc. Symp. Application on Centrifuge Modelling in Geotechnical Design*, pp. 475–481. Balkema, Rotterdam.

Nunez, I.L. and Randolph, M.F. (1984) Tension pile behaviour in clay-centrifuge modelling techniques. *Proc. Symp. Application of Centrifuge Modelling to Geotechnical Design*, pp. 87–102. Balkema, Rotterdam.

Oldham, D.C.E. (1984) Experiments with lateral loading on single piles in sand. *Proc. Symp. Application of Centrifuge Modelling to Geotechnical Design*, pp. 121–141. Balkema, Rotterdam.

Phillips, E. (1869) De l'équilibré des solides élastiques semblables. *C.R. Acad. Sci., Paris*, **68**, 75–79.

Pokrovskii, G.I. and Fiodorov, I.S. (1936) Studies of soil pressures and deformations by means of a centrifuge. *Proc. 1st Int. Conf. Soil Mech. Found. Eng.*, Vol. 1, p. 70.

Pokrovskii, G.I. and Fiodorov, I.S. (1975) Centrifugal model testing in the construction industry. Vols. I, II. English translation by Building Research Establishment Library Translation Service.

Sabagh, S. (1984) Cyclic axial load pile tests in sand. *Proc. Symp. Application of Centrifuge Modelling to Geotechnical Design*, pp. 103–121. Balkema, Rotterdam.

Savvidou, C. (1988) Centrifuge modelling of heat transfer in soil. *Proc. Conf. Centrifuge '88*, pp. 583–591. Balkema, Rotterdam.
Schmidt, R.M. (1988) Centrifuge contributions to cratering technology. In *Centrifuges in Soil Mechanics* (eds W.H. Craig, R.G. James and A.N. Schofield), pp. 199–202. Balkema, Rotterdam.
Schofield, A.N. (1980) Cambridge geotechnical centrifuge operations. *Géotechnique*, **20**, 227–268.
Sekiguchi, H. and Phillips, R. (1991) Generation of water waves in a drum centrifuge. *Proc. Conf. Centrifuge '91*, pp. 343–350. Balkema, Rotterdam.
Ter-Stepanian, G.I. and Goldstein, M.N. (1969) Multistoreyed landslides and the strength of soft clays. *Proc. 7th Int. Conf. Soil Mech. Found. Eng.*, Vol. 2, pp. 693–700.
Townsend, F.C., Tabatabai, H., McVay, M.C., Bloomquist, D. and Giu, J.J. (1988) Centrifugal modelling of buried structures subjected to blast loadings. *Proc. Centrifuge '88, Paris*, pp. 473–479. Balkema, Rotterdam.
Vinson, T.S. (1983) Centrifugal modelling to determine ice/structure/geologic foundation interactive forces and failure mechanisms. *Proc. 7th Int. Conf. Port and Ocean Engineering under Arctic Conditions*, pp. 845–854.

2 Centrifuges in modelling: principles and scale effects

R.N. TAYLOR

2.1 Introduction

Centrifuge model testing represents a major tool available to the geotechnical engineer since it enables the study and analysis of design problems by using geotechnical materials. A centrifuge is essentially a sophisticated load frame on which soil samples can be tested. Analogues to this exist in other branches of civil engineering: the hydraulic press in structural engineering, the wind tunnel in aeronautical engineering, the flume in hydraulic engineering and even the triaxial cell in geotechnical engineering. In all cases, a model is tested and the results are then extrapolated to a prototype situation.

Modelling has a major role to play in geotechnical engineering. Physical modelling is concerned with replicating an event comparable to what might exist in the prototype. The model is often a reduced scale version of the prototype and this is particularly true for centrifuge modelling. The two events should obviously be similar and that similarity needs to be related by appropriate scaling laws. These are very standard in areas such as wind-tunnel testing where dimensionless groups are used to relate events at different scales. They are less familiar to geotechnical engineers, although the time factor in consolidation is just such a dimensionless parameter used to relate small-scale laboratory tests to large-scale field situations.

A special feature of geotechnical modelling is the necessity of reproducing the soil behaviour both in terms of strength and stiffness. In geotechnical engineering there can be a wide range of soil behaviour relevant to a particular problem. There are two principal reasons for this: (i) soils were originally deposited in layers and so it is possible to encounter different soil strata in a site which may affect a particular problem in different ways; and (ii) *in situ* stresses change with depth and it is well known that soil behaviour is a function of stress level and stress history. Clearly, in any generalised successful physical modelling it will be important to replicate these features. It is for the second reason that centrifuge modelling is of major use to the geotechnical engineer. Soil models placed at the end of a centrifuge arm can be accelerated so that they are subjected to an inertial radial acceleration field which, as far as the model is concerned, feels like a gravitational acceleration field but many times stronger than Earth's gravity. Soil held in a model container has a free unstressed upper surface and within the soil body the magnitude of stress increases with depth at a rate related to the soil density and the strength of the acceleration field. If the same soil is used in the model as in the prototype and if a careful model preparation procedure is adopted whereby the model is subjected to a similar stress history ensuring that the packing of the soil particles is replicated, then for the centrifuge model subjected to an inertial acceleration field of N times Earth's gravity the vertical stress at depth h_m will be identical to that in the corresponding prototype at depth h_p where $h_p = Nh_m$. This is the basic scaling law of centrifuge modelling, that stress similarity is achieved at homologous points by accelerating a model of scale N to N times Earth's gravity.

Two key issues in centrifuge modelling are scaling laws and scaling errors. Scaling laws can be derived by making use of dimensional analysis (see, for example, Langhaar, 1951) or from a consideration of the governing differential equations. Both methods will be used here to derive the basic scaling laws relevant to a wide range of models. Centrifuge modelling is often criticised as having significant scaling errors due to the non-uniform acceleration field and also the difficulty of representing sufficient detail of the prototype in a small-scale model. It is clearly important to have a proper appreciation of the limitations of a modelling exercise and some of the more common problems will be considered. Also, the different types of model study and their role in geotechnical engineering will be discussed.

2.2 Scaling laws for quasi-static models

2.2.1 Linear dimensions

As discussed above, the basic scaling law derives from the need to ensure stress similarity between the model and corresponding prototype. If an acceleration of N times Earth's gravity (g) is applied to a material of density ρ, then the vertical stress, σ_v at depth h_m in the model (using subscript m to indicate the model) is given by:

$$\sigma_{vm} = \rho N g h_m \tag{2.1}$$

In the prototype, indicated by subscript p, then:

$$\sigma_{vp} = \rho g h_p \tag{2.2}$$

Thus for $\sigma_{vm} = \sigma_{vp}$, then $h_m = h_p N^{-1}$ and the scale factor (model:prototype) for linear dimensions is $1:N$. Since the model is a linear scale representation of the prototype, then displacements will also have a scale factor of $1:N$. It follows therefore that strains have a scale factor of $1:1$ and so the part of the soil stress–strain curve mobilised in the model will be identical to the prototype. (It should be noted that since displacements are reduced, it is not

usually possible to create residual shear planes with the same residual
strengths as the prototype.)

The Earth's gravity is uniform for the practical range of soil depths
encountered in civil engineering. When using a centrifuge to generate the high
acceleration field required for physical modelling, there is a slight variation in
acceleration through the model. This is because the inertial acceleration field
is given by $\omega^2 r$ where ω is the angular rotational speed of the centrifuge and r
is the radius to any element in the soil model. This apparent problem turns
out to be minor if care is taken to select the radius at which the gravity scale
factor N is determined.

The distributions of vertical stress in the model and corresponding prototype are shown in Figure 2.1. These distributions of vertical stress are
compared directly in Figure 2.2 where they are plotted against corresponding

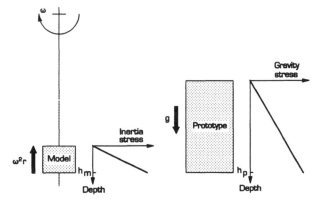

Figure 2.1 Inertial stresses in a centrifuge model induced by rotation about a fixed axis
correspond to gravitational stresses in the corresponding prototype.

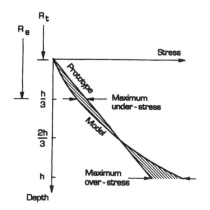

Figure 2.2 Comparison of stress variation with depth in a centrifuge model and its
corresponding prototype.

depth; note that the non-linear variation of stress in the model is shown exaggerated for clarity. In the prototype, the vertical stress at depth $h_p = Nh_m$ is given by:

$$\sigma_{vp} = \rho g h_p = \rho g N h_m \tag{2.3}$$

The scale factor N needs to be calculated at an effective centrifuge radius for the model R_e such that:

$$Ng = \omega^2 R_e \tag{2.4}$$

If the radius to the top of the model is R_t, then the vertical stress at depth z in the model can be determined from:

$$\sigma_{vm} = \int_0^z \rho \omega^2 (R_t + z)\,dz = \rho \omega^2 z \left(R_t + \frac{z}{2} \right) \tag{2.5}$$

If the vertical stress in model and prototype are identical at depth $z = h_i$ (as shown in Figure 2.2) then, from equations (2.3), (2.4) and (2.5), it can be shown that:

$$R_e = R_t + 0.5h_i \tag{2.6}$$

A convenient rule for minimising the error in stress distribution is derived by considering the relative magnitudes of under- and over-stress (see Figure 2.2). The ratio, r_u, of the maximum under-stress, which occurs at model depth $0.5h_i$, to the prototype stress at that depth is given by:

$$r_u = \frac{0.5h_i \rho g N - 0.5h_i \rho \omega^2 \left(R_t + \dfrac{0.5h_i}{2} \right)}{0.5h_i \rho g N} \tag{2.7}$$

When combined with equations (2.4) and (2.6), this reduces to:

$$r_u = \frac{h_i}{4R_e} \tag{2.8}$$

Similarly, the ratio, r_o, of maximum over-stress, which occurs at the base of the model, h_m, to the prototype stress at that depth can be shown to be:

$$r_o = \frac{h_m - h_i}{2R_e} \tag{2.9}$$

Equating the two ratios r_u and r_o gives:

$$h_i = \frac{2}{3} h_m \tag{2.10}$$

and so:

$$r_u = r_o = \frac{h_m}{6R_e} \tag{2.11}$$

Also, using equation (2.6):

$$R_e = R_t + \frac{h_m}{3} \tag{2.12}$$

Using this rule, there is exact correspondence in stress between model and prototype at two-thirds model depth and the effective centrifuge radius should be measured from the central axis to one-third the depth of the model. The maximum error is given by equation (2.11). For most geotechnical centrifuges, h_m/R_e is less than 0.2 and therefore the maximum error in the stress profile is minor and generally less than 3% of the prototype stress. It is important to note that even for relatively small radius centrifuges (say 1.5 m effective radius), the error due to the non-linear stress distribution is quite small for moderately large models of say 300 mm height.

2.2.2 Consolidation (diffusion) and seepage

Consolidation relates to the dissipation of excess pore pressures and is a diffusion event. It is easiest to examine the scaling law for time of consolidation using dimensional analysis. The degree of consolidation is described by the dimensionless time factor T_v defined as:

$$T_v = \frac{c_v t}{H^2} \tag{2.13}$$

where c_v is the coefficient of consolidation, t is time and H is a distance related to a drainage path length. For the same degree of consolidation, T_v will be the same in model and prototype and so:

$$\frac{c_{vm} t_m}{H_m^2} = \frac{c_{vp} t_p}{H_p^2} \tag{2.14}$$

Since $H_p = N H_m$, then:

$$t_m = \frac{1}{N^2} \frac{c_{vp}}{c_{vm}} t_p \tag{2.15}$$

Hence, if the same soil is used in model and prototype (which is usually the case) then the scale factor for time is $1:N^2$. The scale factor would need to be adjusted, as indicated by equation (2.15), if for some reason the coefficients of consolidation were not the same in the model and prototype. Thus a consolidation event lasting 400 days in the prototype can be reproduced in a one-hour centrifuge run at $100\,g$. This scaling also applies to other diffusion events such as heat transfer by conduction.

It is important to recognise that this apparent speeding up of time-related processes is a result of the reduced geometrical scale in the model; the centrifuge is not a time machine. The same sort of scale factor applies to oedometer tests in which a small element of soil is tested in the laboratory to determine its consolidation behaviour. The time taken for the completion of consolidation in the small element is related to the corresponding time in the field by the square of the scale factor which links the drainage path lengths in the two situations.

The scaling laws for time of seepage flow have led to a minor controversy which centres on two issues: the interpretation of hydraulic gradient, and whether or not Darcy's permeability is a fundamental parameter. What is clear is that the rate of seepage flow is increased in a centrifuge model (otherwise spin driers, for example, would be of no practical use). Darcy's law for seepage flow is:

$$v = ki \tag{2.16}$$

where v is the superficial velocity of seepage flow, k, is the coefficient of permeability and i is the hydraulic gradient.

In fluid mechanics and hydrology, the intrinsic permeability, K, is often used. It is defined by:

$$K = \frac{vk}{\rho g} \tag{2.17}$$

where v and ρ are, respectively, the dynamic viscosity and density of the fluid. In this definition, K is a function of the shape, size and packing of the soil grains. Thus, if the same pore fluid is used in model and prototype then Darcy's coefficient of permeability is apparently a function of gravitational acceleration with the implication that $k_m = Nk_p$. Hydraulic gradient is defined as the ratio of a drop in head of pore fluid, Δs, to the length, Δl, over which that drop occurs as shown in Figure 2.3. Thus hydraulic gradient is

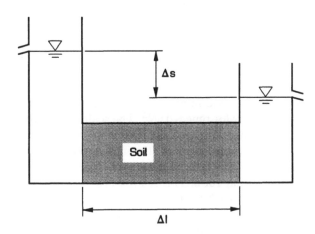

Figure 2.3 Hydraulic gradient defined by a simple problem involving steady seepage flow.

dimensionless and, it is argued, does not scale with acceleration, i.e. $i_m = i_p$. On that basis,

$$v_m = i_m k_m = i_p N k_p = N v_p \qquad (2.18)$$

and thus the velocity of flow is N times greater in the model than the prototype as expected.

While this line of reasoning is logical, it has the worrying implication that soils would become impermeable under a zero gravity field. That is because a tacit assumption has been made that all seepage flow is gravity driven. At zero gravity, porous media would then appear impermeable because there would then be no pressure gradient driving the seepage flow, although there could be an apparent hydraulic gradient. There is therefore some merit in questioning the concept of hydraulic gradient as being simply the ratio of two lengths. Hydraulic gradient is more realistically a ratio of pressure drop over a distance and is representative of a pressure gradient. Since stresses (or pressures) are the same and distances are reduced N-fold from prototype to model, clearly this interpretation of hydraulic gradient implies $i_m = N i_p$. Darcy's permeability can then be treated as a material parameter, i.e. $k_m = k_p$ and thus:

$$v_m = i_m k_m = N i_p k_p = N v_p \qquad (2.19)$$

i.e. seepage velocity has a scale factor of $1:N$; the same result as determined in equation (2.18).

The flow paths along which pore fluid travels have a scale factor for length of $1:N$. The time for seepage flow is then:

$$t_m = \frac{L_m}{v_m} = \left(\frac{L_p}{N}\right)\left(\frac{1}{N v_p}\right) = \frac{1}{N^2} t_p \qquad (2.20)$$

and so the scale factor for time in seepage flow problems is $1:N^2$, the same as determined for diffusion and consolidation events. If for some reason the soils in model and prototype have different permeabilities, it can be shown that the scaling relationship for time becomes:

$$t_m = \frac{1}{N^2} \frac{k_p}{k_m} t_p \qquad (2.21)$$

Hence the effect of the different permeabilities can be taken into account.

In all the above, it has been assumed that the soil is fully saturated. While this is usually a good assumption, there are circumstances when the problem being modelled will involve flow in partially saturated soils. This is an important issue particularly in the context of near surface pollution migration following a spillage and preliminary investigations into the scaling have been reported by Goforth et al. (1991) and Cooke and Mitchell (1991). It is likely that this will feature more and more in future centrifuge studies.

2.3 Scaling laws for dynamic models

Dynamic events such as earthquake loading or cratering require special consideration in order to define appropriate scaling laws. For such problems, it is simplest to consider the basic differential equation describing the cyclic motion x_p in the prototype (which is small compared to the overall dimensions):

$$x_p = a_p \sin(2\pi f_p t_p) \qquad (2.22)$$

where a_p is the amplitude of the motion of frequency f_p. Differentiating equation (2.22) gives:

$$\frac{dx_p}{dt_p} = 2\pi f_p a_p \cos(2\pi f_p t_p); \quad \text{velocity magnitude } 2\pi f_p a_p \qquad (2.23)$$

$$\frac{d^2 x_p}{dt_p^2} = -(2\pi f_p)^2 a_p \sin(2\pi f_p t_p); \quad \text{acceleration magnitude } (2\pi f_p)^2 a_p \quad (2.24)$$

Using an analogous expression for motion in the model, the following expressions can be derived:

displacement magnitude: a_m
velocity magnitude: $(2\pi f_m)a_m$
acceleration magnitude: $(2\pi f_m)^2 a_m$

In the model, linear dimensions and accelerations have scale factors $1:N$ and $1:N^{-1}$, respectively, in order to retain similarity. From the above, it is clear that this can be achieved if $a_m = N^{-1} a_p$ and $f_m = N f_p$. An important consequence of this is that the velocity magnitude will then be the same in the model and the prototype, i.e. velocity has a scale factor $1:1$. The time scaling factor for dynamic events is therefore $1:N$ in contrast to the $1:N^2$ time scale factor for diffusion or seepage events. With these scale factors, it can be seen that 10 cycles of a 1 Hz earthquake (duration 10 s) with amplitude 0.1 m can be represented by a centrifuge model tested at $100\,g$ subjected to 10 cycles of a 100 Hz earthquake (duration 0.1 s) having an amplitude of 1 mm.

The conflict in time scaling factors requires special consideration. For example, in modelling the stability of clay embankments subjected to earthquake loading, it could be argued that during the earthquake, the low permeability of the soil would prevent significant flow of water. Since there is practically no seepage flow or diffusion of water, the dynamic time scaling factor of $1:N$ should apply. Subsequent to the earthquake, any dissipation of excess pore pressures would be modelled using a time scaling factor of $1:N^2$. However, a problem arises in the study of liquefaction of saturated fine sands during an earthquake where excess pore pressure dissipation will occur during the earthquake event. In that case, it is necessary to ensure that the time scaling factor for motion is the same as that for fluid flow. The approach usually adopted is to decrease the effective permeability of the soil by increasing the viscosity of the pore fluid (see equation 2.17). For example, a 100 cSt silicone fluid is 100 times more viscous than water but has virtually the same density. Darcy's permeability will therefore appear to be 100 times less for a sand saturated with the silicone fluid compared to the same sand saturated with water. Thus a centrifuge model using sand saturated with silicone fluid and tested at 100 g would have a time scale factor for fluid flow of 1:100 which is the same as the time scale factor for dynamic motion given by 1:N since N is 100 in this example.

2.4 Scale effects

In physical modelling studies, it is seldom possible to replicate precisely all details of the prototype and some approximations have to be made. It is important to recognise that model studies are not perfect and to inquire into the nature of any shortcomings, often referred to as scale effects, and to evaluate their magnitude. The influence of the non-uniform acceleration field created in centrifuge models is an example of a scale effect and was considered in section 2.2.1. The following are some examples of scale effects; others may be relevant in any particular centrifuge study and it is the responsibility of the centrifuge worker to establish for any particular project the extent to which model test results can be extrapolated to a prototype scale.

A good technique for checking for scale effects is the 'modelling of models'. It is particularly useful when no prototype is available for verifying the model test results. Centrifuge models of different scale are tested at appropriate accelerations such that they then correspond to the same prototype. The models should predict the same behaviour and thus provide a useful internal check on the modelling procedure. However, it should be noted that the scale range of the models is usually limited. For example, it might be possible to model an event at 40 g and 120 g on the same centrifuge. This is only a range of scale of 3 compared to the scale of 40 scale needed to extrapolate from the larger model to the prototype. Thus, while 'modelling of models' provides a valuable internal check on the modelling procedure, it does not in itself provide a guarantee that model data can be extrapolated successfully to the prototype scale.

2.4.1 Particle size effects

The most common question asked of centrifuge workers is how can centrifuge modelling be justified if the soil particles are not reduced in size by a factor of N. In increasing the model scale to the prototype in the mind's eye, it might appear sensible to also increase the particle size. Thus a fine sand used in a 1:100 scale model might be thought of as representing a gravel. But by the same argument, a clay would then be thought of as representing a fine sand. This argument is clearly flawed since a clay has very different stress–strain characteristics to a fine sand. There could be a problem if an attempt was made to model at high acceleration and hence at very small scale an event in a prototype soil consisting mainly of a coarse soil (gravel). In that case, the soil grain size would be significant when compared to model dimensions and it is unlikely that the model would mobilise the same stress–strain curve in the soil as would be the case in the prototype. Local effects of the soil grains would influence the behaviour rather than the soil appearing like a continuum as would be the case in the prototype.

It is therefore sensible to develop simple guidelines on the critical ratio between a major dimension in the model to the average grain diameter to avoid problems of particle size effects. This was the approach adopted by Ovesen (1979, 1985) who investigated the performance of circular foundations on sand by undertaking a series of experiments using different sized models at different accelerations such that they corresponded to the same prototype. The data were generally internally consistent which validated the centrifuge technique, though it was noted that there was some deviation from the common behaviour when the ratio of foundation diameter to grain size was less than about 15. Thus the particle size scale effect could be quantified to some extent. However, this approach may be too simplistic and in some cases it may be necessary to consider the ratio of particle size to shear band width (Tatsuoka et al., 1991). The important point is to recognise that in some circumstances particle size effects may be important and the model test series should include sufficient relevant investigation to assess its significance in the problem being studied.

2.4.2   Rotational acceleration field

While a centrifuge is an extremely convenient method of generating an artificial high gravitational acceleration field, problems are created by the rotation about a fixed axis. The inertial radial acceleration is proportional to the radius which leads to a variation with depth in the model; this was discussed in section 2.2.1. Also, this acceleration is directed towards the centre of rotation and hence in the horizontal plane, there is a change in its direction relative to vertical across the width of the model. There is therefore a lateral component of acceleration, the effect of which needs to be recognised. For a model having a half width of 200 mm and an effective radius of 1.6 m, this lateral acceleration has a maximum value of 2/16 or 0.125 times the 'vertical' acceleration. This can be quite significant if there is a major area of activity near a side wall of the model container. For some centrifuges, the major vertical plane lies in the horizontal plane of rotation and if the radius of the centrifuge is relatively small, some centrifuge workers have adopted the practice of shaping models to take account of the radial nature of the acceleration field. Alternatively, it is good practice to ensure that major events occur in the central region of the model where the error due to the radial nature of the acceleration field is small.

Another problem caused by generating the acceleration field by rotation is the Coriolis acceleration which is developed when there is movement of the model in the plane of rotation. This might be the horizontal movement of base shaking earthquake simulation on models whose major vertical plane lies parallel to the plane of rotation. In trying to avoid this, many centrifuges now arrange for the major vertical plane of the model to be perpendicular to the plane of rotation. However, there may be vertical velocities in the plane of rotation and the effect of Coriolis accelerations needs to be assessed. The following gives guidelines for the range of velocities in a model for which Coriolis effects could be considered negligible. The Coriolis acceleration a_c is related to the angular velocity, ω, of the centrifuge and the velocity, v, of a mass within the model as:

$$a_c = 2\omega v \qquad (2.25)$$

The inertial acceleration, a, of the model is:

$$a = \omega^2 R_e = \omega V \qquad (2.26)$$

where V is the velocity of the model in centrifuge flight. It is generally assumed that Coriolis effects would be negligible if the ratio a_c/a was less than 10% which implies $v < 0.05V$ (see section 7.2). This gives an upper limit on v for relatively slow events.

At the other extreme, for example, the high velocity of soil ejected during blast simulation, it was argued that the radius of curvature, r_c, of the path followed by a moving mass in the model should not be less than the effective radius of the centrifuge. The Coriolis acceleration can then be written as:

$$a_c = 2\omega v = \frac{v^2}{r_c}$$

i.e.

$$r_c = \frac{v}{2\omega} \qquad (2.27)$$

Since $V = \omega R_e$, then for $r_c > R_e$, $v > 2V$. Thus it is concluded that the range of velocities within a model which would not lead to significant Coriolis effects is given by:

$$0.05V > v > 2V \qquad (2.28)$$

Further discussion of Coriolis effects is presented in chapter 7.

2.4.3   Construction effects

Geotechnical engineering is often concerned with the effects of construction and this can pose many difficulties for the centrifuge worker. It is very difficult to excavate or build during centrifuge flight. The soil is very heavy and any model equipment needs to be small, lightweight and very strong and usually requires skilful design. Though there are many difficulties, new techniques and devices are being developed and these are often reported at speciality centrifuge conferences.

In modelling construction or installation processes, the first thoughts need to be directed towards defining the essential details which have to be modelled and, importantly, those details which are of secondary importance and can be taken into account in some approximate way. Even if approximations are made, centrifuge data will still be useful as these can be taken into account in any back-analysis, so verifying the analysis for later application to the prototype event. An example is modelling embankment construction on soft clay where the foundation behaviour is the main feature under investigation rather than the actual embankment. Many centrifuge centres have developed hoppers which can be used to build up an embankment made of dry sand during centrifuge flight rather than use the proposed embankment material or construction procedure. Nevertheless, it is still possible to study the foundation strata under this loading and any changes in behaviour when, for example, ground improvement techniques are used.

Craig (1983) considered construction effects in the context of modelling pile foundations. If the performance of the piles under lateral loading is being investigated, then it would be reasonable to adopt the easy approach of installing the piles prior to starting the centrifuge. Although the stress distribution due to installation is not correctly modelled, this is a minor effect on the overall performance. However, if the performance of the piles under axial loading is to be studied it is essential to install the piles during centrifuge flight since the load capacity is critically dependent on the lateral stresses developed during installation.

2.5 Model tests

Centrifuge testing concerns physical modelling of geotechnical events. However, it is not restricted to studying a particular prototype with a view to improving design but rather it is a means by which the general understanding of geotechnical events and processes can be better understood. Centrifuge test series can be designed with different objectives in mind; the following categories of model tests are similar to those identified by James (1972).

● The study of a particular problem (for example an embankment) for which there are some difficult design decisions to be made. In such a study, there is clearly a need to replicate sufficient of the essential features of the prototype so that the model test data can be extrapolated to the prototype scale and so give a sensible assessment of its behaviour.

- The study of a general problem with no particular prototype in mind. The investigation is directed at making general statements about a particular class of problem, for example the long-term stability of retaining walls or the patterns of settlement caused by tunnel construction. Each model is a prototype in its own right and the results of a series of tests can be correlated by making good use of dimensional analysis. The purpose of using the centrifuge is then to generate realistic stress distributions so that overall the model test data can be sensibly applied to field situations and can also assist in developing new analyses.
- The detailed study of stress changes and displacements relevant to a particular class of problem. The purpose of such tests is to gain information on soil behaviour which can then assist with developing constitutive models and so improving analysis.

Of these categories, the second is the most applicable to the majority of centrifuge studies. This is, in part, because most centrifuge facilities are established in research institutions and a significant proportion of the centrifuge income comes from research-based programmes. It is pleasing to note that good progress has been made in Japan in using centrifuge tests for more routine design work.

Many of the early programmes of centrifuge studies were essentially investigations of mechanisms of collapse. The centrifuge is particularly useful in these studies since there is proper replication of self-weight effects and realistic failures can be observed, particularly if two-dimensional models are tested with viewing of a vertical section through the Perspex side wall of a model container. There is special merit in centrifuge tests since real soil is used with proper modelling of behaviour. Therefore mechanisms developed in the model will be realistic rather than predetermined as in other forms of analysis. The scope of tests is now much more advanced and serviceability as well as collapse can be studied. This is particularly useful since in geotechnical engineering details of pre-failure patterns of deformation are often just as valuable as factors of safety against failure. Also, centrifuge studies now extend well beyond the traditional problems and centrifuge projects have included studies on burial of heat-generating waste (Maddocks and Savvidou, 1984), stability of iron ore concentrates cargoes during shipping (Atkinson and Taylor, 1994) and the generation of ice-floe forces on offshore structures.

In many problems, as well as studying mechanisms of collapse it is important to assess the relative effects of key parameters, often related to geometry, which influence the mechanism of deformation or collapse. Parametric studies can be successfully undertaken using a centrifuge since it is possible to have good control over the soil models. In this way, it is possible to determine the relative significance of certain parameters which may not be possible by other forms of analysis.

Model studies can, and should, be treated as prototypes in their own right. They embody details directly comparable to prototypes, in particular the correct distribution of stress and stress–strain behaviour. Therefore, they can give excellent data for validation of numerical codes and analysis. These range from predictions of collapse to detailed modelling of deformations. The latter is particularly difficult numerically due to the problems of realistic representation of the small strain behaviour of soils. However, recent research is encouraging and centrifuge test data have been found to provide useful comparisons with detailed finite-element calculations. Also, centrifuge data can be useful even if, for some reason, the models do not fully replicate a particular prototype situation. The numerical code can still be validated against the model test data and that code can then be used for analysis of the full-scale prototype taking into account its special features and boundary conditions.

Site-specific studies are the most difficult type of test to undertake and are the least common. Usually, major decisions are required on what details should be included in the model and on how best to incorporate natural variations across a site within the small-scale model. In studies of the Oosterschelde storm surge barrier, Craig (1984) commented on the importance of soft-sand pockets in the foundation sea-bed soils. These occurred randomly in the field but in the model it was decided to account for them by including a number of evenly spaced sand inclusions occupying a fixed percentage of the total volume of foundation soil. In this way it was possible to quantify to some extent the influence of the sand pockets.

Modelling of specific sites requires the recovery of field samples. This could be in the form of intact blocks which are then trimmed to size and loaded in centrifuge containers for subsequent reconsolidation on the centrifuge and testing. Alternatively, the site soil could be reconstituted and consolidated such that the profile of effective stress history in the model corresponded to the prototype. Reproducing the consolidation history is not especially difficult using consolidation presses with a downward hydraulic gradient consolidation facility. However, effects of ageing and some degree of cementing at grain contacts may have an important influence on behaviour and should be recreated. A promising technique for reproducing ageing effects by consolidation at elevated temperatures is described by Tsuchida et al. (1991).

If an intact block is recovered for later modelling of a site, some thought needs to be given as to the reconsolidation process such that the profile of stress history in the model properly represents the site. If the block is recovered from near the top of the strata of interest and reconsolidated on the centrifuge using a surcharge load to represent any soil eroded from the site in geological time, then all elements of the model with depth will experience the same maximum pre-consolidation stress profile as the *in situ* site. The centrifuge can then be stopped, the surcharge removed and the sample then reconstituted in centrifuge flight. In this way, the prototype effective stress profile and effective stress history will be correctly represented in the model. (It should be noted that this would not be the case for a sample recovered from the bottom of the strata.) However, any 'structure' or 'fabric' present in the field sample is likely to be destroyed by this process and some attempt at ageing of the sample may be necessary if all aspects of the prototype behaviour are to be replicated.

2.6 Summary

The general background to centrifuge modelling has been presented. This has involved derivation of the most fundamental scaling laws using standard methods based either on dimensional analysis or the governing differential equations. In any modelling study, it is important to investigate sources of potential error and for centrifuge modelling the most commonly identified errors have been considered and shown, in general, to be of minor significance. Centrifuge model testing is a technique of increasing relevance to engineering design and practice and some examples of the types of study undertaken have been presented. The increasingly wide variety of centrifuge studies to engineering applications can be examined by reference to the specialist international conferences.

References

Atkinson J.H. and Taylor, R.N. (1994) Drainage and stability of iron ore concentrate cargoes. *Centrifuge '94. Singapore*, pp. 417–422. Balkema, Rotterdam.

Cooke, A.B. and Mitchell, R.J. (1991) Evaluation of contaminant transport in partially saturated soils. *Centrifuge '91, Boulder, Colorado*, pp. 503–508. Balkema, Rotterdam.

Craig, W.H. (1983) Simulation of foundations for offshore structures using centrifuge modelling. In *Developments in Geotechnical Engineering* (ed. P.K. Banerjee), pp. 1–27. Applied Science Publishers, Barking.

Craig, W.H. (1984) Centrifuge modelling for site-specific prototypes. *Symp. Application of Centrifuge Modelling to Geotechnical Design, University of Manchester*, pp. 473–489. Balkema, Rotterdam.

Fuglsang, L.D. and Ovesen, N.K. (1988) The application of the theory of modelling to centrifuge studies. In *Centrifuges in Soil Mechanics* (eds W.H. Craig, R.G. James and A.N. Schofield), pp. 119–138. Balkema, Rotterdam.

Goforth, G.F., Townsend, F.C. and Bloomquist, D. (1991) Saturated and unsaturated fluid flow in a centrifuge. *Centrifuge '91, Boulder, Colorado*, pp. 497–502. Balkema, Rotterdam.

James, R.G. (1972) Some aspects of soil mechanics model testing. In *Stress–Strain Behaviour of Soil, Proc. Roscoe Mem. Symp., Foulis*, pp. 417–440.

Langhaar, H.L. (1951) *Dimensional Analysis and Theory of Models*. John Wiley, New York.

Maddocks, D.V. and Savvidou, C. (1984) The effects of heat transfer from a hot penetrator installed in the ocean bed. *Symp. Application of Centrifuge Modelling to Geotechnical Design, University of Manchester*, pp. 336–355. Balkema, Rotterdam.

Ovesen, N.K. (1979) The scaling law relationship—Panel Discussion. *Proc. 7th Eur. Conf. Soil Mech. Found. Eng., Brighton*, No. 4, pp. 319–323.

Tatsuoka, F., Okahara, M., Tanaka, T., Tani, K., Morimoto, T. and Siddiquee, M.S.A. (1991) Progressive failure and particle size effect in bearing capacity of a footing in sand. *ASCE Geotechnical Engineering Congress 1991*, Vol. II (Geotechnical Special Publication 27), pp. 788–802.

Tsuchida, T., Kobayashi, M. and Mizukami, J. (1991) Effect of ageing of marine clay and its duplication by high temperature consolidation. *Soils and Found.*, **31** (4), 133–147.

3 Centrifuge modelling: practical considerations
R. PHILLIPS

3.1 Introduction

As centrifuge modelling is an experimental science, there are practical considerations to be made in the design and conduct of any centrifuge model test. The primary objective of the model test is to obtain high-quality and reliable data of a physical event. The model test should therefore be based on experience using proven apparatus and techniques. For centrifuge modelling to advance, gradual development of these apparatus and techniques is required.

The model test is a multi-disciplinary activity, generally involving mechanical, hydraulic, electronic and control engineering as well as geotechnics. The test objectives are constrained by these activities. Some of these constraints are becoming less severe with the implementation of new technology. The centrifuge model test is also constrained by more mundane restrictions such as resource, payload capacity, increased self-weight and communication to the remote centrifuge environment.

The art of centrifuge modelling is to minimise the effects of these constraints while maximising the quality of the geotechnical data obtained. This chapter will assist the reader in understanding some of the practical considerations and review some of the apparatus and techniques currently available to the centrifuge modeller.

Centrifuge modelling has been used extensively for geotechnical studies, and is also being applied to more general civil engineering studies including rock mechanics, hydraulics, structures and cold regions. Most of the considerations presented in this chapter are also applicable to these research areas.

3.2 Geotechnical centrifuges

There are many different types of centrifuges, used for example in material processing, aeronautics and motion simulation. The geotechnical centrifuge is characterised by its rugged nature, large payload capability and low-speed operation. Geotechnical centrifuges can be sub-divided into two main classes: beam centrifuges and drum centrifuges.

The beam centrifuge generally comprises a central spindle supporting a pair of parallel arms which hold the platform on which the test package is placed. The radial force acting on the central spindle should be minimised as shown in equation (3.1):

$$|\Sigma m \mathbf{r} \omega^2| \rightarrow 0 \qquad (3.1)$$

where m is the centre of mass of a component at a vector $\mathbf{r}$ from the rotor axis and ω is the rotational speed of the centrifuge. This equation can be simplified to:

$$|\Sigma m \mathbf{r}| \rightarrow 0 \qquad (3.2)$$

From equation (3.2) the centrifuge can be viewed as a simple mass balance with the central rotor replaced by a central point support. Balance must be considered in two orthogonal directions. Traditionally balance has been achieved by symmetry; such beam centrifuges are called balanced-beam centrifuges. In recent years, beam centrifuge design has changed and masses at large radii, such as the platform and test package, are counterbalanced by a larger mass at a smaller radius. This new generation of beam centrifuges, such as the one at Delft Geotechnics and those manufactured by Acutronic, are more efficient as the aerodynamic drag on the rotor assembly is reduced.

Beam centrifuges traditionally rotate in a horizontal plane. The acceleration field acting on the model is the resultant of the centrifuge acceleration field and the Earth's gravitational field. The behaviour of the model will depend on the orientation of the model on the centrifuge platform to this resultant acceleration field. Beam centrifuges can be subdivided into three platform types (fixed, restrained and swinging) as depicted in Figure 3.1.

On the fixed platform the test package is attached to the vertical face plate. On the centre-line, the resultant acceleration is always effectively inclined to the platform at $n{:}1$, where n is the centrifuge acceleration at the platform. When the centrifuge is started or stopped, restraints are necessary to retain the soil and low shear modulus materials, such as fluids, in the test package.

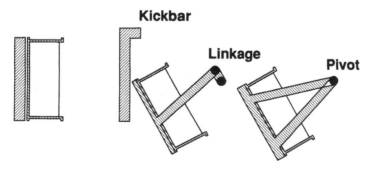

Figure 3.1 Types of beam centrifuge platform: left, fixed; centre, restrained; right, swinging.

Restraint techniques have been successfully developed at the University of Manchester and elsewhere using flexible membranes in conjunction with a vaccum. These restraints need to be relaxed during the centrifuge test to minimise their effect on the model behaviour.

Restrained platforms are used, for example, on the Cambridge University and ISMES beam centrifuges. This platform type eliminates the need for a low g retention system. At rest, the swinging platform surface is slightly inclined to the horizontal. As the centrifuge speed increases the platform swings up into the vertical plane at about $10\,g$ where it is restrained by a kick bar. Further speed increases cause the increased weight of platform to twist the support linkage so that the platform seats against, and is carried by, the strong cradle. The minimum inclination of the resultant acceleration to the platform surface is about 10:1 (84°) during the swing up process. Above $10\,g$, the resultant inclination increases to about $n:1$ to the platform surface.

The effect of this inclination at the test acceleration level on the soil model can be eliminated by either sloping the soil strata within the test package or more simply by placing a wedge between the test package and the platform surface. These measures are only fully effective at the test acceleration level.

Swinging platforms were used in the original Russian centrifuges (Pokrovskii and Fiodorov, 1953). Swinging platforms have been adopted in the new generation of Acutronic civil engineering centrifuges and elsewhere. The platform surface will always be normal to the resultant acceleration, provided the hinge is frictionless and the platform symmetrically loaded. The effects of hinge friction have been discussed by Xuedoon (1988). There is a slight variation of vertical stress across a horizontal plane as the model surface is not parallel to the axis of rotation. A major benefit of these platforms is the ease of access to the user.

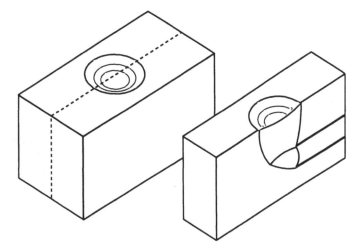

Figure 3.2 Movement seen around tunnel heading. Left, full model; right, half model.

Drum centrifuges are being increasingly used for geotechnical research. Some of the first drum centrifuge tests are described by Schofield (1978). The drum centrifuge can be considered as having a fixed platform and it is not easy to retain models in the drum. Models can be constructed in the drum during rotation due to the continuous circumferential extent of the soil sample. Pore suction can be used to hold these models in place while the centrifuge is stopped to instrument the model and place any actuators. Some drum centrifuges rotate in a vertical plane, and the radial acceleration component on the model then has a cyclic component due to the Earth's gravity. This cyclic component can be used to create water movement in model tests. The Earth's gravity is beneficial during model making when operating in this mode. The new mini-drum centrifuge under development at Cambridge University has the ability to rotate from a horizontal to vertical plane during centrifuge rotation.

3.3 Containers

In the drum centrifuge the model is normally contained within the drum itself. In beam centrifuges, the model must be safely held within a container. The geotechnical centrifuge model test is normally a simulation of the behaviour of an infinite half space with a localised perturbation. The container boundaries should replicate the behaviour of the far field half-space.

In models of static events, this generally requires creating one-dimensional consolidation boundary conditions using ideally frictionless vertical walls of high lateral stiffness to prevent significant lateral soil displacement. For modelling dynamic events, particularly earthquakes, the boundary conditions are more complex with prevention of energy reflections while maintaining the correct dynamic shear stiffness and permitting the evolution of complementary shear stresses (Schofield and Zeng, 1992). Other boundary condition requirements may include thermal control for cold-regions' problems and hydraulic control for environmental problems. Boundary conditions at the top and bottom of the model must also be considered.

The half space of interest may include vertical planes of symmetry. These planes can be replaced by low-friction stiff surfaces, such as the side-walls of the container walls to reduce the size of the scale model. Sub-surface deformation patterns along these planes can be visualised if these frictionless surfaces are transparent, as depicted in Figure 3.2.

Practically, the container walls will be frictional. The coefficient of friction can be estimated from direct shearbox tests of the boundary conditions. For clay tests, side-wall friction has been reduced by lubricating the smoothly coated stiff walls with a water-resistant grease. For sand testing, a hard glass sheet can be placed between the sand mass and the wall. For small strain problems in sand, the frictional resistance can be further decreased using lubricated latex sheets at the boundaries. However, the frictional resistance may increase since membrane tensions can become appreciable as boundary shear displacements increase.

For plane strain models, the model could be very narrow. Since side-wall friction is always present to some extent, plane strain models should be sufficiently wide so that side-wall friction is not a significant proportion of the resisting forces. The effect of side-wall friction can be further reduced by taking measurements along the centre-line of the model. Similar considerations are required in one-dimensional experiments and large strain events, such as consolidation.

Containers for two- and three-dimensional studies should be about twice as long as the soil depth they contain to minimise boundary effects. The tops of the container walls are useful for providing support for actuators and interfaces, and as a datum surface during model making. For static testing, the lateral displacement of the container wall should be less than about 0.1% of the retained height of soil to have minimal effect on lateral earth pressures.

In the centrifuge model simulation vertical planes are mapped into the centrifuge cylindrical coordinate system (r,θ,z) as radial planes. During large strain events, such as consolidation, the arc length between these radial planes lengthens with increasing radius which would cause lateral straining of the soil sample. To maintain an overall condition of zero lateral strain, the arc length $(r\delta\theta)$ should be kept constant. This condition is reasonably approximated by having the container side-walls parallel.

For general use, circular containers, or tubs, are very versatile due to their inherent lateral stiffness and consequent light mass. The tubs retain the maximum soil plan area with the minimum boundary material and are relatively inexpensive to construct. Also, tubs can be easily sealed to retain pressure.

Rectangular containers, or boxes, are more massive and more expensive to construct than a tub of the same carrying capacity and stiffness. However, since centrifuge platforms are usually rectangular, boxes permit the available soil plan area to be maximised. Transparent side-walls can be used in the box to permit visualisation of sub-surface events. Sometimes, these boxes are not stiff enough to retain the soil pressures during the preparation of over-consolidated samples. This difficulty can be circumvented by consolidating the soil into a liner and then transferring the soil and the liner into the box. Care must be taken to ensure that the internal dimensions of the con-solidometer and the box are identical, otherwise lateral straining of the soil sample will occur and the lateral earth pressures will change. Similar effects will occur when using bulkheads within the box to retain the soil mass.

The containers should not leak. The container boundaries may require glanded ports for instrumentation and service channels. The base of the container should be stiff enough to prevent significant disturbance of the sample when the base is unsupported during model preparation and mechanical handling, and the base should be flat to be evenly supported by the centrifuge platform. The base of the container should also accommodate selective base drainage of the soil sample as required.

3.4 Test design

A geotechnical centrifuge test is normally designed to model a generic prototype situation. As with many other reduced-scale modelling techniques, such as hydraulic modelling, not every aspect of the prototype behaviour can be correctly modelled. Attention must be made to model directly those factors which are expected to dictate the prototype behaviour, such as the effective stress conditions in the soil. For those factors which cannot be directly modelled, the modeller must still ensure that the correct class of behaviour is simulated if possible. A commonly occurring example is modelling the flow of pore fluid through the soil skeleton. If the same pore fluid and soil are used in the model and the prototype then the Reynolds number is larger by the scaling factor, n in the model. Laminar flow conditions in the model soil skeleton can still be maintained by ensuring that the Reynolds number in the model is less than unity (Bear, 1972).

Although the model may not be an exact replica of the prototype, it is still a unique physical event. In prototype terms, the engineer can view the model as 'the site next door' where conditions are very similar if not identical to their own.

In the centrifuge model only those processes which are dominated by gravitational effects will be automatically enhanced. To verify the effects that these processes have on the prototype behaviour, the technique of 'modelling of models' can be employed as described by Schofield (1980). Similarities between the different models can be attributed to these processes. Differences in behaviour can assist in separating the effects of different processes: Miyake et al. (1988) modelled the process of soft-clay sedimentation and consolidation and separated the two effects by modelling of models techniques.

The principle of 'modelling of models' was discussed by (for example) Ko (1988) and is demonstrated in Figure 3.3: the same 10 m high prototype could be modelled at full scale, at 1/10th scale, or at 1/100th scale at points A1, A2 and A3, respectively. Normally the range of scales used for modelling of models is narrower than indicated and does not include full-scale tests; more care is then required to extrapolate the results to prototype scale. The effects of stress and size must also be considered when comparing tests (Ko, 1988).

For small-size models the effect of particle size is also important. Soil is a particulate medium. Modellers, for example Fugslang and Ovesen (1988), have found that at least 30 particles must be in contact with each linear dimension of the model structure for the observed behaviour to be representative of the prototype behaviour. Care must be taken before scaling down the

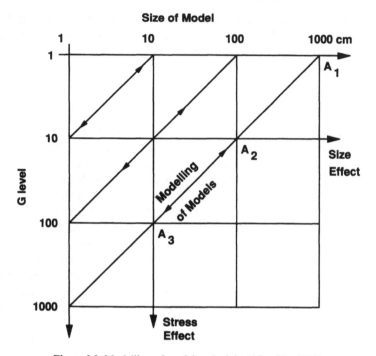

Figure 3.3 Modelling of models principle. (After Ko, 1988).

particle size in the model to ensure that the mechanical properties of the particles are not changed, including their angularity and crushing strength, as demonstrated by Bolton and Lau (1988).

The geometric scale factor for the model is selected to fit the prototype situation under study into the model container with minimal boundary effects. The choice of scale factor will be constrained by the maximum model size, which is related to the payload capacity of the centrifuge, and the operational domain of the centrifuge. In general, the scale factor should be as small as possible to maximise the size of the model: small models are more difficult to instrument and more sensitive to the presence of the instrumentation and the model making procedure. Small models of simple boundary value problems are, however, valuable for performing parametric studies with multiple models in one soil sample.

Some prototype situations may be too large for direct centrifuge modelling. For deep problems, the effective stress levels in the soil can be increased by downward seepage (Zelikson, 1969). This technique was used by Nunez and Randolph (1984) in centrifuge model tests of long piles.

The appropriate centrifuge acceleration level is normally identical to the geometric scaling factor, but may be different when equivalent materials and partial similarity are required as described by Craig (1993). The centrifuge acceleration level is not constant but increases linearly with centrifuge radius. Schofield (1980) showed that the appropriate centrifuge acceleration level should be selected at one-third the depth of interest in the soil model, and that provided the overall soil depth did not exceed 10% of the effective centrifuge radius the error in assuming that this acceleration level is constant with depth is tolerable.

Ideally soil strata within the soil model should be formed at the same curvature as that of the centrifuge. The majority of centrifuge model tests are conducted with level strata. If the package width is about 20% of the centrifuge radius in the circumferential direction, then a lateral acceleration of 10% of the centrifuge acceleration will be induced at the outside edge of the level strata. This effect can be minimised by placing the area of interest along the centre-line of the strongbox.

The time scaling factors are determined from the appropriate centrifuge acceleration level, N, using the scaling laws. From these time factors the actuation frequency of the soil model can be determined. The method of actuation and required actuation power can then be selected.

In many cases, the actuation frequency required to directly model diffusion events is too fast for the available means of actuation or may require too much power. The higher actuation frequency may also induce inertial effects that are not present in the prototype. The actuation frequency must then be reduced to more manageable levels. This reduced actuation frequency must not cause a significant change in the dissipation of pore pressure occurring during the actuation.

Where significant pore pressure changes would occur in granular materials, the modeller can choose to increase the viscosity of the pore fluid to retard pore pressure dissipation. Increasing pore fluid viscosity by the scaling factor is common when modelling dynamic events in sand to match the time scaling factors for inertia and diffusion. The modeller must also ensure that changing the pore fluid does not significantly affect the mechanical behaviour of the soil medium. Wilson (1988) has shown that damping within the soil medium is increased close to resonance when viscous fluids are used.

The instrumentation for monitoring the model test can be selected knowing the type and expected range of measurands and the required monitoring frequency. The selection of modelling materials, actuators and instrumentation are discussed in sections 3.5 to 3.8.

The model structure is normally scaled to have the same external geometry as the prototype. Other scaling requirements might include the bearing stress, the stiffness and the strength of the structure relative to the soil medium. Frequently, these model structures are manufactured from different materials than the prototype to satisfy the selected scaling criteria.

The model test must be integrated with the centrifuge. The design should include how communication is established with the test package and how the model will be affected by the centrifuge environment. As the centrifuge rotates, most of the power required to rotate the centrifuge is dissipated in aerodynamic drag creating heat and a potential increase of temperature within the centrifuge chamber. For tight temperature control within the model test, the test package may have to be enclosed within thermal control barriers. The heat may be partly dissipated by ventilation of the rotating mass of air within the chamber. Air movements may cause undesirable effects to the exposed model such as buffeting and high evaporation rates which should be controlled by protecting the exposed model.

3.5 Model preparation

The most important aspect of a geotechnical soil model is the effective stress profile. The effective stress history, the current effective stress state and the effective stress path followed during the test will dictate the behaviour of the model.

Centifuge model tests can be performed on undisturbed soil samples, if the effective stress conditions in the sample are representative of the prototype. Macro-fabric present in the undisturbed model sample, such as structure, fissures, inclusions and potential drainage paths, may not scale to be representative of the conditions in the prototype.

Macro-fabric present in undisturbed soil samples can be eliminated by remoulding of the natural soils. The centrifuge model is then constructed from these remoulded materials. A site investigation of the prototype situation is required to reconstruct a representative centrifuge model. The use of remoulded soil distorts the materials history including the effect of ageing. Techniques for artificially ageing remoulded small soil samples are being developed (Tsuchida et al., 1991). Without such techniques the strength of the undisturbed material may not be modelled correctly in the remoulded sample. The remoulded sample should then be treated as an equivalent material and allowance made in the test design for the change in soil strength and behaviour.

For more generic conditions, reconstituted laboratory soils are normally used with well-defined soil properties. The behaviour of these laboratory soils can be altered to produce the required material behaviour by the use of mixtures (for example, Kimura et al., 1991). The soil conditions in such models are well controlled, permitting numerical analyses to be validated against such physical model test data.

Remoulded granular soil models can be prepared by tamping and pluviation techniques. The soil models are generally too large to be compacted on vibrating tables. Tamped samples can be prepared moist or dry for most grain size distributions. The sample is placed in layers which are then compacted by tamping to achieve the required overall density. There may be a variation of density within the tamped layers.

Dry pluviation techniques can be used for uniformly graded dry sands. Finer silt material become air borne and will not pluviate. Air borne dust is an important consideration using the pluviation technique. The dust should be contained in the model preparation area and the modeller should wear a face mask to prevent silicosis.

The density of pluviated samples can be accurately controlled by the energy imparted to the sand particles: dense samples are created by pouring the sand slowly from a height whereas loose samples are created by slumping the sand quickly into the model container. These and other factors controlling pluviated sand sample densities are presented by Eid (1988). Samples pluviated into a rotating-drum centrifuge may have a lateral velocity relative to the soil surface; this will cause densification of the sample. The pluviation technique should be carefully chosen to reduce cross-anisotropy within the pluviated samples. Single-point hoppers are particularly useful when creating highly-instrumented samples of complex geometry. The dry sand surface can be accurately shaped using a vacuum system. Mechanical shaping of the sand surface will disrupt the surface density of the sample.

Saturated samples can be created by pluviating the sand through the pore fluid. This technique can only be used to create lightly instrumented, relatively loose samples of simple geometry. These samples may not have an acceptable degree of saturation. Pluviated samples are better saturated after construction. The simplest technique is to introduce the pore fluid from a header tank through a base drainage layer into the dry model. The driving head should be kept below the hydrostatic head necessary to fluidise the sand sample. Movement of the saturation front should be controlled to prevent air pockets becoming trapped within the saturated material.

Higher degrees of saturation can be achieved by evacuating air from the sand sample before the pore fluid is introduced. The degree of saturation for water-saturated samples can be further increased by replacing the air within the sample by carbon dioxide. The carbon dioxide can be introduced after evacuating the air from the sample. The inlet pressure of the gas must not be sufficient to cause fluidisation of the sand sample.

It is advisable when saturating samples under vacuum to have the pore fluid in the header tank under the same vacuum as the sand sample. This has the advantages of de-airing the pore fluid before it is introduced into the sample and of minimising the pressure differential between the header tank and the soil sample. Boiling of the pore fluid under vacuum should be avoided.

Some pore fluids may be too viscous to penetrate the soil matrix. The viscosity of some of these fluids may reduce sufficiently at elevated temperatures to enter the soil matrix. Saturation using such fluids will require the whole soil sample and header tank to be heated using a water bath or similar.

Remoulded clay and silt samples can be created by tamping. For better-defined stress histories, clay and silt samples should be reconstituted from a slurry. The slurry should be mixed at about twice the liquid limit of the material. De-ionised water can be used as the pore fluid to minimise chemical effects and bacterial growth within the sample. The slurry should be mixed under vacuum for about two hours to de-air the slurry and create a smooth slurry. The resulting slurry is then placed into a consolidometer. Care must be taken not to trap air pockets within the slurry mass and drainage layers during placement which would decrease the high degree of saturation of the slurry mass. Layered samples can be created if slurry ingress is prevented into the underlying layers.

Silt and clay slurries can be consolidated in the centrifuge. Caution is needed to prevent the generation of high pore pressures within the slurry mass which may cause piping through the clay mass and preferential drainage paths. The high pore pressures can be avoided by accelerating the centrifuge in stages to the required test speed with delays at each stage to allow dissipation of excess pore pressures. As the shear strength of the sample increases during consolidation, the surface of the sample will hold a particular inclination. Therefore, for centrifuges with restrained platforms, the resultant surface may not be correctly inclined at the required test speed. Consolidation of deep clay layers in the centrifuge is a lengthy procedure.

Silt and clay samples are more often formed in a consolidometer, which may also be the test container. The initial consolidation increment should be about 5–10 kPa, unless measures have been taken to prevent extrusion of the slurry from the consolidometer. After this first increment, the consolidation pressure can be increased after 80% of primary consolidation is achieved. The consolidation is easily monitored by measuring the vertical settlement of the consolidometer piston or by measuring the amount of pore fluid expelled from the sample. The consolidation pressure can be successively doubled until the required maximum consolidation pressure is achieved.

This consolidation technique is normally used to create a uniform consolidation pressure with depth. Trapezoidal variations of consolidation pressure can be created by inducing seepage within the consolidometer. Normally the effective consolidation pressure at the surface of the sample is required to be lower than that at the base. The consolidation pressure required at the base of the sample is applied at the surface of the sample and the base of the sample drained to atmosphere. The applied consolidation pressure can be imposed by two methods. A sealed impermeable piston can be used to apply the full base consolidation pressure which is reduced effectively at the clay surface by the pore pressure acting in the drainage layer between the sample and the piston. The second method is to use an unsealed piston which applies the effective surface consolidation pressure. The base consolidation pressure is applied by pressurising the pore fluid around the piston. This consolidation technique has been called downward hydraulic gradient consolidation (Zelikson, 1969). Using this technique a number of different consolidation profiles can be successively applied to the clay sample to reproduce most pre-consolidation profiles, including normally consolidated clay. By reversing the technique and applying fluid pressure at the base of the sample, upward hydraulic gradient consolidation can also be undertaken, to create, for example, over-consolidated surface crusts.

Consolidometers should be designed so that lateral displacement of the consolidometer wall should be less than 0.1% of the retained height of soil to maintain earth pressures in their at rest condition as indicated by Yamaguchi *et al.* (1976).

After primary consolidation is complete at the maximum pre-consolidation profile, the sample can be unloaded in the consolidometer. After each stage of unloading, the pore pressures within the soil sample are in suction. If these suctions are too high, cavitation with a consequential loss of strength may occur within the soil mass. In the presence of excess pore fluid, these suctions will dissipate in a controlled manner.

The height of the sample should be measured immediately before and after it is unloaded in the consolidometer. The elastic heave of the sample can then be assessed and accommodated during construction of the model. After the sample is unloaded, the effective stresses within the sample are retained by pore suctions. These suctions can be maintained by removing excess pore fluid from around the sample and preventing air entry into the sample. The sample can be sealed with plastic cling film or similar to reduce pore pressure dissipation.

For thick clay samples, the time required to establish full-equilibrium effective-stress conditions during the centrifuge test may be excessive. This reconsolidation time can be reduced by decreasing the drainage path lengths within the soil model. These shorter paths can be accomplished by creating thin granular drainage layers within the soil sample when it is first constructed. After consolidation, these drainage layers must be connected to an external source of pore fluid at the correct potential. After unloading from the consolidometer, these layers will be in suction. Care must be taken when connecting into these layers to prevent air entry or the layers may become air-locked and ineffective. Radial drainage can also be created using vertical wick-wells formed from washed wool or string.

Another option is to create a quasi-drainage boundary in the soil sample: the clay sample is unloaded in the consolidometer to a uniform consolidation pressure which is the average effective stress applied to the clay sample during the centrifuge test. During the centrifuge test the upper part of the sample will swell and absorb water and the lower part of the sample will consolidate and expel water, thus creating a quasi-drainage boundary within the clay sample.

The preferred option to minimise consolidation time in the centrifuge is to finally consolidate the clay in the consolidometer close to the effective stress profile it will experience during the centrifuge test using downward hydraulic gradient consolidation. The surface over-consolidation ratio must not exceed 10 or cracking of the clay surface may occur disrupting the seepage flow

(Kusakabe, 1982). The minimum amount of time should be taken between unloading the consolidometer to starting the centrifuge test. The effective stress within the clay sample can be monitored using pore-pressure transducers and the model sample actuated when an acceptable effective stress profile is restored.

Reconstituted kaolin powder has been used extensively to create clay centrifuge models. Kaolin is a coarse grained clay with a relatively high permeability which consolidates rapidly minimising model preparation and centrifuge test durations. Mixtures of clay, silt and sand offer the centrifuge modeller a wider range of material behaviour and properties than those available from a single material type.

Other centrifuge modelling materials have included equivalent materials, particularly for rock mechanics studies, and photoelastic media (Clark, 1988).

Commercially available bulk materials, such as sand, silt and clay powder, are frequently used as modelling materials. The mechanical properties of these materials may change with time. Index properties of these modelling materials should be routinely measured to ensure consistency of the material. These materials are sometimes recycled after a centrifuge test and re-used. These recycled materials should be tested to ensure there has been no significant degradation of the material.

3.6 Fluid control

Fluid control in the model test is important for maintaining the correct drainage and effective stress conditions. Water is most commonly used as the test fluid.

As the centrifuge speed changes, free fluid surfaces will flow to become normal to the resultant acceleration field. This movement of fluid across the sample surface may cause erosion or over-topping in the sample. These effects can be reduced by either submerging the whole soil surface or by limiting the amount of fluid on the surface during speed changes. When testing cohesive samples, the presence of free fluid during speed changes will allow dissipation of pore suctions within the sample. This dissipation can be reduced by adding free fluid to the test sample after the centrifuge has started, and removing the free fluid before the centrifuge is stopped. These changes in fluid mass within the test package must be controlled to prevent excessive out-of-balance forces developing on the centrifuge rotor.

If the soil surface is not submerged, excessive evaporation may lead to drying and desiccation of the soil surface. This evaporation can be controlled by covering the surface with a protective coating such as liquid paraffin, which will minimise surface evaporation without significantly affecting the behaviour of the soil. Near-surface pore suctions, from evaporation and capillary action, will increase the effective stress in the soil skeleton and should be monitored and controlled to acceptable levels. Surface pore suctions can be used to create surface crusts.

The fluid level within the sample should be monitored and controlled. Fluid may be lost from the sample due to evaporation or leakage from the package. The fluid level can be controlled using standpipes. The standpipes are external to the test package and connected with pipework to the drainage layers within the soil sample. The pressure heads in the pipework are increased under the centrifuge acceleration. Care is required to prevent cavitation within the pipework from flow under these increased heads.

The required equipotential level in the standpipe and elsewhere in the test package can be calculated allowing for centrifuge curvature and the Earth's gravity effects. This equipotential level can either be controlled using a fixed overflow or a control system. The fixed overflow is simple to implement but will not maintain a fixed level of fluid relative to the sample during consolidation of the sample. A simple control system may comprise of a fluid-level indicator and a dosing pump: as the fluid level in the package decreases the dosing pump restores the fluid level to the required level.

The fixed overflow system can be used as a constant loss system, where fluid is fed continuously into the test package and overflowed to waste. Generally, fluid overflows such as water do not need to be retained in the package but can be dumped into the centrifuge chamber. Care should be taken to pipe these overflows away from any electrical devices. Other fluids, such as oil and chemicals, must be safely retained within the test package. Fluids can be fed into the standpipe through hydraulic slip-rings. The passage of fluid down the centrifuge rotor to the test package is analogous to fluid flowing down into a steep-sided valley. The standpipe then serves as a stilling tank to dissipate the energy in the fluid and provide the fluid at the correct potential for the model. Fluid feeds piped directly from the slip-rings into the model package may cause significant erosion of the soil sample.

The standpipes can be used to change the fluid control level during the course of the test. The elevation of pipework from the standpipe or the discharge point into the test package must not exceed the standpipe control level or there will be no fluid flow. The fluid control level can be changed by selection of a different overflow level using valves, or by changing the control level.

3.7   Actuation

An important consideration in a centrifuge model test design is how to actuate the model to simulate the prototype perturbation. Typically, the model actuation frequency should be at least the scaling factor, N, times faster than the prototype to simulate similar degrees of pore-pressure dissipation. The actuator's power density can be defined as the power per unit volume of actuator. Dimensional analysis shows that the actuator power density required for the model should then be at least N times greater than the prototype power density for correct scaling.

This increase in power density is not normally achievable, and the model actuator is proportionally either larger than the prototype or not as powerful. Similarly, instrumentation, interfaces and other devices are not scaled. The modeller would prefer the majority of the payload to be the soil model, but the actuator and other devices must be accommodated.

The payload capacity occupied by the actuator and other devices can be reduced by placing sub-systems away from the test package on either the centrifuge rotor or off the centrifuge. Such sub-systems might include power supplies, power amplifiers, controllers and conditioning modules. This separation can be advantageous as the sub-systems are subjected to a lower acceleration field and can occupy a larger volume than can be accommodated at the test package. Communication, however, needs to be established between the sub-systems and the test package. The bandwidth and number of communication channels required may sometimes limit the performance of the actuator. For example, long hydraulic hoses between a hydraulic power pack and a hydraulic cylinder will limit the dynamic response of the system. There are many electrical lines between a brushless servo-motor and controller which may make this connection inefficient through slip-rings.

The mode of actuation should be kept simple and resource effective. Frequently, the centrifuge model test objectives will evolve as a test programme proceeds requiring changes to the actuator. Simple actuators are likely to be more compact and more reliable than complex actuators. In recent years, actuators have become simpler as motion controllers have become more sophisticated. This development has been very beneficial to centrifuge modelling permitting a range of complex tasks to be accomplished using a combination of simple actuators, as demonstrated by McVay et al. (1994).

The actuator should not restrict the behaviour of the model. For example, if model tests are being conducted on vertical bearing capacity, the footing should be free to rotate and translate under the vertical load. If the footing is rigidly attached to the actuator then these motions will be prevented. Allowance should be made for flexure of actuator and support systems under load. Such flexure may cause undesirable loading of the soil model and may inhibit stiffness measurements of the model.

Actuators can be developed from commercially available products, or specially constructed. There is a wide range of standard commercial products suitable for use in centrifuge modelling. The modeller should consider using such products before developing their own: the centrifuge model test research objective is, after all, in geotechnics not electrical, control or mechanical engineering!

When selecting a commercial product, the principle of operation and the physical construction of the product must be considered to determine how the product will perform on the centrifuge. Most manufacturers and suppliers do not warrant their products for operation under high-acceleration fields, but are usually interested to provide technical support for this unusual application. Most commercial products are sized for continuous operation over a number of years. Such products may be over-designed for the small number of duty cycles required in the centrifuge model test, permitting smaller-sized products to be used.

Some customised actuator parts, especially interfaces between standard components and the model, for a centrifuge model test may be required. Designs for these parts can be sought from other experienced centrifuge modellers, from researchers in other fields with analogous requirements, or from specialist design companies. The experience of other centrifuge modellers is available through texts (such as this one), the proceedings of speciality conferences, the theses and reports published by the many geotechnical centrifuge centres worldwide and by discussion with the experienced modellers at these centres. The modeller is encouraged, where possible, to standardise their modelling techniques and equipment with those already developed.

Actuators can be powered from different sources including electricity, hydraulics, pneumatics and latent energy, as described in the following paragraphs. It is impractical to store sufficient electrical, hydraulic and pneumatic power on the centrifuge to meet the total demand of the model test. Pneumatic and hydraulic power can be generated on the centrifuge using electrical power. Some electrical, hydraulic and pneumatic slip-rings are therefore required on a centrifuge, even though more data acquisition and control communications are being established using non-contact techniques such as optical slip-rings and radio local area networks (LANs), rather than mechanical slip-rings. Limited quantities of electrical, hydraulic and pneumatic power can be stored on the centrifuge, using for example batteries, accumulators and gas cylinders to meet peak demands, such as earthquake actuation. Latent energy is easily stored on the centrifuge as potential energy in the actuator, kinetic energy in the centrifuge rotor or in explosives.

Electrical power is normally transmitted as an alternating current which may radiate electrical noise. This noise pick-up can be reduced by the use of shielded twisted pair cables for the electrical power, proper attention to earthing and electrical connections, and physical separation of the power and data cables and slip-rings.

Hydraulic fluid, such as water or oil, will increase in weight under the centrifuge acceleration. Hydraulic power systems must accommodate these hydrostatic pressure increases. Water can be used for low-pressure applications mixed with a rust-inhibitor and lubricant. Water is denser and less viscous than hydraulic oil, and therefore can exert much higher hydrostatic pressures with lower pressure losses. Hydraulic oil will be required for high-pressure applications. Most hydraulic systems are designed to leak. This leakage must be collected against the hydrostatic back-pressure on the return line. Hydraulic slip-rings are also required for fluid feeds to the centrifuge model, such as surface water control as described above, and for other fluids such as refrigerant.

Low-pressure pneumatic systems to about 15 bar can be driven through pneumatic slip-rings. These slip-rings should be lubricated and cooled to extend the slip-ring seal life. High pneumatic pressures to about 200 bar can be stored in compressed gas cylinders on the rotor. Pneumatic pressure is excellent, as the increased gas weight is negligible, for force-controlled loading. Displacement-controlled monotonic loading can be achieved using a pneumatic cylinder: one chamber of the pneumatic piston is connected to the pneumatic pressure supply and the other chamber is filled with hydraulic fluid, which is vented through a control orifice. When pneumatic pressure is applied, movement of the piston is restricted by the flow rate through the orifice.

The centrifuge acceleration field can usefully be exploited as a power source. Component self-weight has been used for the installation and compressive loading of piles, suction caissons, penetrometers and footings. Tensile loading can be imparted from the buoyancy of a float in water. The self-weight of water can also be used as a loading source. Excavation events have been simulated by the removal of dense fluid from a retaining pressure bag. The inorganic salt zinc chloride is particularly effective because of its high specific gravity and exceptional solubility in water. Zinc chloride is inexpensive, but also very corrosive (see section 4.1.3).

The 'bumpy road' earthquake actuator utilises a fraction of the kinetic energy stored in the Cambridge University beam centrifuge as its power source. A medium-sized centrifuge will store about 10 MJ of energy in its rotor. Controlled explosions are also a very useful energy source. Access to these two energy sources may impart excessive loads to the centrifuge.

The weight of all components within the test package will be carried in compression into the platform, probably through the container. In general, support systems are simpler if the components are carried in compression. All components should be firmly attached to the test package to prevent disturbance to the model and to permit mechanical handling of a complete package. The orientation of actuators with respect to the acceleration field is important to ensure correct operation of the actuator under its increased self-weight.

The increased weight of pistons in hydraulic cylinders and pneumatic cylinders may need to be supported by pressure if the cylinder is mounted vertically, or may cause the piston to rack or leak in the cylinder if the cylinder is mounted horizontally. If the required support pressure is too high in the former case, it can be reduced by decreasing the weight of the piston and piston rod: these items could be re-made in aluminium alloy. The piston rod could be replaced by a smaller rod or hollow tube depending on the axial load requirements. Cylinders can be acquired with a double piston rod, extending from both ends of the cylinder. The piston rod can then be fitted with guides to prevent racking of the piston within the cylinder. The additional piston rod is also useful as a reference from which to measure displacement of the piston.

The orientation of shuttle valves also needs to be considered. Normally, the shuttles should be mounted vertically, with the weight of the shuttle keeping the valve down in its normal operating position. The weight of the spindle needs to be considered when selecting the valve actuator. For example, for solenoid or pneumatically actuated valves, the spindle weight can be calculated in terms of an actuation pressure knowing the geometry of the valve. The specified pressure rating for the actuator is then the required pressure for the test plus the actuation pressure.

Electric motors are normally best orientated with their main rotor axis in line with the centrifuge acceleration field. The increased rotor weight can then be carried through thrust bearings attached to the motor output shaft. The rotor will flex when placed across the acceleration field. This flexure and the radial play in the rotor supports may permit the rotor and stator to short. The type of electric motor and controller to use is dependent on the power, speed and control requirements of the model test. Generally, three-phase electric motors are preferable to single-phase motors as they have a higher power-to-frame size ratio and radiate less electrical noise. Permanent magnet motors are beneficial due to their simple construction and reduced number of electrical connections. Brushless servo-motors have an excellent power-to-frame size ratio and a simple construction, but require a sophisticated electronic controller and power amplifier.

Solid state electronics, such as printed circuit boards, can be used successfully in the centrifuge. In general, printed circuit boards should either be aligned radially with the acceleration field, or supported on foam rubber or similar to prevent flexure of the board and breakage of solder tracks and components. The behaviour of large or delicate components should be considered under increased self-weight. Large components such as transformers and heat sinks may require additional mechanical support. Delicate components may need to be potted for support. Electrolytic capacitors have been found to distort and change capacitance under their increased weight. Electrical connections should be aligned to mate better in the centrifuge and not be distressed.

3.8   Instrumentation

Centrifuge model test behaviour can be monitored by a variety of instrumentation. Available instrumentation includes not only a wide range of transducers but also visual techniques as described below. New instrumentation is being developed which may be applicable to centrifuge model testing. As this instrumentation becomes available, its suitability can be assessed using the guidelines presented below. Particularly useful instrumentation is anticipated from the areas of remote sensing and fibre optics.

Transducers in contact with the centrifuge model should be small and rugged enough to resist not only their increased self-weight but also mechanical handling during test preparation and disassembly. Solid-state transducers are particularly suitable. The operating principle of the transducer must be considered. Normally, the transducer is required to be capable of continuous monitoring throughout the centrifuge test, such as pressure transducers. More infrequent monitoring may be acceptable such as deformations before and after an event. For continuous monitoring, the transducer should have an adequate frequency response, which is normally one or two orders of magnitude higher than that required in the prototype.

The transducer output may require conditioning to be interfaced to the data acquisition system. The transducers and conditioning modules will be limited by the space associated with the test package. The transducer should be reliable.

Transducers embedded within the soil model should be miniature with dimensions of about 10 mm. These buried transducers may act as ground anchors. The model test must be designed such that it does not become a test of reinforced earth. The transducers should be orientated to minimise reinforcement effects. The transducer leads should be flexible and run orthogonal to the direction of principal movement. Transducer lead runs should also minimise potential drainage path effects. Buried transducers and their leads must withstand the high ambient pressure levels within the soil mass and the high ambient water pressures, when used in saturated media.

Pore-pressure transducers are fitted with a porous element to isolate the fluid pressure for measurement. The movement of fluid through this porous element will mechanically filter the frequency response of the transducer (Lee, 1991). The response is dependent on the degree of saturation and porosity of this element. Push-fit elements are recommended for these transducers. These elements can then be de-aired in pore fluid and fitted to the transducer under pore fluid to ensure a high degree of saturation of the transducer. The elements should be replaced if they become blocked. Ceramic elements are recommended for use in clay and coarser elements, such as sintered bronze, for use in granular soils. These transducers can be used without porous elements for use in standpipes and free pore fluid.

Commercially available Druck PDCR81 transducers are commonly used for pore-pressure measurement. There are other commercially available pore-pressure transducers, but these are generally larger than those supplied by Druck. Pressure transducers smaller than those supplied by Druck are commercially available but these transducers are not suitable for burial in a soil model. The PDCR81 transducer is a differential pressure transducer. The reference pressure is provided through the hollow electrical lead. The integrity of this air passage must be ensured. Calibration and installation procedures for these transducers are described by König et al. (1994).

Total stress transducers are required to define more completely the state of stress at the boundaries or within the soil model. The stiffness of the transducer is very important. For boundary-stress measurements, the stiffness of the transducer should be similar to the boundary stiffness for a representative stress measurement. Fluid-filled diaphragm transducers are suitable, but they only measure normal force. Normal and shear forces can be quantified using the Stroud cell, but this cell is not as stiff as those above. Soil arching also affects pore pressure measurement (Kutter et al., 1988).

Displacements can be measured with potentiometers or linearly variable differential transformers. Both these transducers require contact with the model. For vertical measurements, the spindle weight may need to be carried on a pad to prevent indentation of the spindle into the model. For horizontal measurements, the spindle may require mechanical assistance, such as a spring or glue, to maintain the spindle in contact with the model. Potentiometers should be orientated to keep the wipers in contact with the resistive elements.

Temperature measurements can be made using thermocouples or thermistors. Boundary temperature measurements can be made cheaply using a digital thermometer in the view of a CCD camera.

Some instrumentation is test specific and requires development. Such instrumentation may be load cells, which are also the linkage between the actuator and the model structure, or the structures themselves such as instrumented piles, retaining walls or geotextiles. This instrumentation is normally designed to measure strain using foil strain gauges. If possible, the strain gauges should be configured in a complete Wheatstone bridge to reduce thermal effects within the instrument. These thermal effects can be minimised by correct selection of the strain gauges and by restricting the power dissipation within each gauge.

Strain gauges for instruments required for only two or three of months can be bonded with super glue, as long-term instrument stability is not required. In these model structures, the lead wires will be proportionally large. The support and routing of these lead wires is an important consideration in the design of the instrument. The strain gauges and lead wires will need to be protected and sealed when used on embedded instrumentation. The completed instrument should be exercised and calibrated over its working load range before the centrifuge test. The instrument should be load cycled about 20 times to reduce hysteresis within the instrument.

Surface cracking has been sensed using conductive paint or thin foil. The crack is sensed from the break in electrical continuity. Other customised instrumentation has included miniature resistivity probes at Cambridge University and University of Western Australia (Hensley and Savvidou,

1993), wave height gauges at Cambridge University (Phillips and Sekiguchi, 1991) and radiation sensors (Zimmie et al., 1993).

Many of the electronic instruments mentioned only provide detailed point measurements within the model. Visual measurements provide an overview of the model behaviour. Movements of the soil model can be observed by placing markers within the soil mass. In sands these markers can be thin coloured sand bands. In clay these markers can be noodles or lead threads. These markers are accurately placed during model preparation. Exposure of these markers after the centrifuge test reveals the plastic deformation of the soil model. The sand bands can be exposed in vertical sections: in fine sand, pore suctions are sufficient to hold a vertical face created by dissecting the sample with a vacuum cleaner. In coarser sands, the pore fluid can be replaced by sugar water, which under heat will cement the sand structure together providing stability during dissection.

Lead threads are formed by injecting a suspension of lead powder in water-soluble cutting oil into the soil sample to leave a lead-coated shaft. (These shafts are potential drainage paths in the model.) Radiographic examination of these lead threads reveal distortion within the model such as rupture band formation. Sufficient time must be left for the oil to diffuse into the soil sample before the centrifuge test, otherwise the lead thread shafts may hydrofracture under the pressure of the heavy lead suspension. Lead threads can be injected into pre-drilled holes in granular models. Lead-impregnated, home-made pasta noodles can also be used.

Vectors of face movements can be determined by indenting markers into the soil face. Some of these markers may be lead shot to define the model boundaries during radiographic examination. Generally, these markers should be small and about the same density as the soil they displace to minimise their influence on the model behaviour. Markers used behind a transparent window should have a low frictional resistance against the window in order to track the soil movements.

Marker positions can be tracked from successive photographic negatives of the model in rotation or television cameras. Stereoscopic cameras mounted very close to the central axis of the centrifuge have been used at the University of Manchester to map surface topography of centrifuge models. Photographic cameras can also be mounted in the centrifuge containment structure. High-resolution photographs are taken of the model in rotation using a short-duration high-intensity flash system synchronised to the rotation of the centrifuge. Flash durations are typically $5\,\mu s$. Measurements of successive polyester-based photographic negatives permit the marker positions to be tracked to an accuracy of about 0.1 mm in a 500 mm wide field of view. Strains within the plane of movement can be assessed by finite differentiation of these movements.

Markers can also be tracked using image analysis systems as described by Garnier et al. (1991) and Allersma (1991). For accurate measurements, these images should be stored digitally rather than in analogue form to minimise distortion of the image on the storage medium. For small centrifuges, stroboscopes or CCD cameras can be synchronised to the centrifuge rotation to view the centrifuge model. Small CCD cameras are now frequently used on the test package to monitor various aspects of the centrifuge model test. Inaccessible locations for the CCD camera can be visualised using mirrors or endoscopes fixed to the CCD camera. Topographic mapping of soil surfaces after the centrifuge test also provide useful information.

3.9 Data acquisition

A typical outline of a data-acquisition system is shown in Figure 3.4. Most of the modules required for such a system, including the data acquisition and control software, are commercially available. The suppliers handbooks for such systems provide a free and very reasonable introduction to data acquisition.

The requirements of centrifuge data-acquisition systems are unusual. The system is required to record multi-channel data about two orders of magnitude faster than in the prototype. The system is also required to be flexible to accommodate the acquisition requirements of many different types of centrifuge test. Such a system may typically be required to acquire data from 16

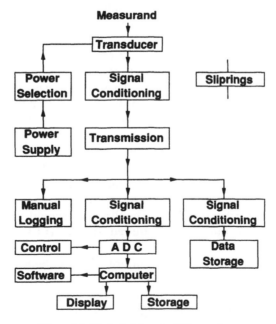

Figure 3.4 Typical data-acquisition system.

transducers at 10 kHz per channel for a second during an earthquake simulation and from 50 transducers at 0.01 Hz per channel for two days during a pollution-migration experiment. Centrifuge model tests can be relatively expensive one-off tests, so redundancy in the data acquisition system to ensure data have been captured is advisable.

In the centrifuge environment, failures of the data-acquisition system do occasionally occur. The system architecture should be modular to permit faults to be easily traced and rectified. This modularity also aids the evolution, expansion and upgrading of the system.

The possible positions of the slip-rings in the data-acquisitions system (Figure 3.4) has changed dramatically over the last 20 years. In many of the original centrifuges, the transducers were connected directly to the slip-rings and the rest of an analogue data-acquisition system was located off the centrifuge. With the progress of micro-electronics and digital technology, it is possible to have the complete data-aquisition system on the test package with remote digital communication to the centrifuge modeller.

The slip-rings do provide a bottleneck on the system, due to the number, bandwidth and capacity of the slip-rings available. Some of these constraints are being bypassed with the use of optical slip-rings and radio-LAN communication, or by multiplexing.

In general, all the transducer signals should be conditioned close to source to provide high-level signals in a common format for onward transmission. The signal conditioning may include decade amplification and filtering of the transducer signals. The signal conditioning and power selection for the transducers can be placed in an interface box attached to the test package. This interface box can provide a very tidy connection between the multiplicity of transducers and the centrifuge rotor wiring. Connectors on the transducer leads and rotor wiring should be strain-relieved so that the cable weights are not carried by the electrical connections. These interface boxes permit pre-test calibration and verification of the transducer performance, and maximum machine utilisation. Transducers monitoring the speed of the centrifuge or the acceleration level at a known radius linked to the data-acquisition system are strongly recommended.

All transducer cables and transmission cables must be adequately supported along their length to carry their increased weight. Electrical-power and signal-transmission cables should be physically separated and properly shielded to reduce electrical pick-up. The slip-rings should be correctly wired for the types of data transmission required, such as IEEE, serial, digital or analogue.

Manual logging is an important feature of the data-acquisition system. Transducer reliability is not high in centrifuge model tests. Manual logging allows the individual transducer outputs to be verified, independent of the computerised data-logging system, against the predicted response from the transducer's calibration. The manual logging also provides a continuous monitoring capability of selected channels. This capability could be used by the centrifuge operator to ensure the integrity of the test package, such as leakage, or provide feedback to a closed-loop control system without compromising the main data-logging system performance.

The main data-logging system will typically include an analogue-to-digital convertor (ADC) interfaced to a computing system. The front-end signal conditioning may be required to optimise the inputs to the ADC; this conditioning may include binary gain amplification, offsetting and filtering. The ADC may include an input multiplexor and a single convertor. The effect of skew between successive inputs must then be considered.

Many complete proven data logging systems are commercially available. The modeller is advised to use one of these systems, rather than develop their own. Modellers should also beware of becoming a test site for the most up-to-date commercial data-acquisition products. Customised systems will be required for specific tests but should only be developed if essential: hardware and software development is time-consuming.

Centrifuge model tests are resource intensive. Each model test is unique, yielding a wealth of data, which if successful is unlikely to be repeated. Modellers should consider backing up their main data-acquisition system with a mass data storage system, such as magnetic tape or optical disc. In the event of problems with the main system or unexpected events, the mass data storage system can be used to recover essential data. The mass data storage system should sample at higher data rates than the main system to ensure that the details of unexpected events can be defined. In an analogue recording system, a magnetic tape recorder provides an excellent mass data-storage system. Digtal mass storage systems are becoming more readily available.

The data obtained from the acquisition software should be suitable for input to the data processing, analysis and reporting software to streamline the procedure of model test reporting. Centrifuge model tests are probably better defined and as highly instrumented as any field trial. Data reduction rates are, however, not accelerated because the data were acquired from a centrifuge test! The modeller should allow sufficient time to fully interrogate their data set and develop a better understanding of their problem.

3.10 Test conduct

The previous sections have considered mainly the design and construction of centrifuge models. Conduct of centrifuge model tests is important. The majority of effort is required in preparation for a centrifuge model test. To benefit from this effort, the modeller is encouraged to prepare a full checklist of all activities required to successfully complete the test. This checklist is best prepared by the modeller carefully thinking through every step of the test:

from the design, through equipment construction, technique development, model-making, package assembly, integration with the centrifuge, package verification, conduct of the test, post-test investigation through to data processing and reporting. The test is a multi-disciplinary project with many different factors to consider. As many test sub-systems as possible should be verified before the centrifuge is started.

The actual centrifuge test is a culmination of effort which may prove exhausting to the modeller especially during extended centrifuge runs. The checklist prompts the user, for example, to turn on the data-acquisition system before actuating the model! The modeller is in control of the model test, and should take time to cross-check their actions to ensure the success of their test.

Safe conduct of every centrifuge model test is the concern of every centrifuge modeller. The safety of the centrifuge must always have a higher priority than the successful completion of any model test. The cost of the centrifuge alone is more expensive than the centrifuge model tests. Centrifuges are very powerful and should be treated with respect. Personnel training in the safe operation of the centrifuge is important.

Every centrifuge model test should be discussed and planned with an experienced centrifuge modeller. Calculations of the stress conditions in the test package and the balance of the centrifuge are strongly advised. Test package configurations change frequently. Centrifuge tests to prove the integrity of the package beyond its normal working condition in the presence of an experienced modeller are advisable.

Centrifuge models are normally constructed away from the centrifuge. Mechanical handling of the test package onto the centrifuge should not cause undue disturbance to the soil model. Sensitive soil models can be transported under vacuum to increase the effective stresses within the model. Very sensitive models can be formed in the centrifuge chamber, but care should be taken not to contaminate the chamber with soil, which may shot blast the centrifuge and associated systems during rotation. Every centrifuge centre should develop a code of practice for centrifuge operations; an example of that used at Cambridge University was presented by Schofield (1980).

When starting and stopping the centrifuge, the angular acceleration should be kept low to prevent the inducement of significant lateral force on the model test. The action is required to prevent swirl of free surface water in drum centrifuges. Consolidation of the test sample to the required effective stress condition is well monitored by the integrated effect of surface settlement rather than individual pore-pressure measurements.

Shortly after the centrifuge run and investigation of the model behaviour, the centrifuge test package should be dismantled and cleaned and the test components safely stored. Most centrifuge facilities are shared by a number of different centrifuge modellers and are in a continual state of flux. Each modeller must be responsible for their own test.

3.11 Conclusion

Centrifuge model testing is a challenging and exciting experimental science. It is a tool for the geotechnical engineer. Like any tool, the success or otherwise of the model test will reflect the effort and aptitude of the modeller. The practical considerations presented in this chapter are intended to assist the modeller to select the correct tools for the job.

References

Allersma, H.G.B. (1991) Using image processing in centrifuge research. *Centrifuge '91* (eds H.Y. Ko and F.G. McLean), pp. 551–558. Balkema, Rotterdam.

Bear, J. (1972) *Dynamics of Fluids in Porous Media.* American Elsevier, New York.

Bolton, M.D. and Lau, C.K. (1988) Scale effects arising from particle size. *Centrifuge '88* (ed. J.F. Corté), pp. 127–134. Balkema, Rotterdam.

Clark, G.B. (1988) Centrifugal testing in rock mechanics. In *Centrifuges in Soil Mechanics* (eds W.H. Craig, R.G. James and A.N. Schofield), pp. 187–198. Balkema, Rotterdam.

Craig, W.H. (1993) Partial similarity in centrifuge models of offshore platforms. *Proc. 4th Canadian Conf. Marine Geotechnical Engineering,* St Johns, Newfoundland, Vol. 3, pp. 1044–1061. C-CORE, Memorial University of Newfoundland.

Eid, W.K. (1988) Scaling Effect in Cone Penetration Testing in Sand. Doctoral thesis, Faculty of Engineering, Virginia Polytechnic Institute and State University.

Fuglsang, L.D. and Ovesen, N.K. (1988) The application of the theory of modelling to centrifuge studies. In *Centrifuges in Soil Mechanics* (eds W.H. Craig, R.G. James and A.N. Schofield), pp. 119–138. Balkema, Rotterdam.

Garnier, J., Chambon, P., Ranaivoson, D., Charrier, J. and Mathurin, R. (1991) Computer image processing for displacement measurement. *Centrifuge '91* (eds H.Y. Ko and F.G. McLean), pp. 543–550. Balkema, Rotterdam.

Hensley, P.J. and Savvidou, C. (1993) Modelling coupled heat and contaminant transport in groundwater. *Int. J. Numerical Anal. Methods Geomech.,* **17**, 493–527.

Kimura, T., Takemura, J., Suemasa, N. and Hiro-oka, A. (1991) Failure of fills due to rain fall. *Centrifuge '91* (eds H.Y. Ko and F.G. McLean), pp. 509–516. Balkema, Rotterdam.

Ko, H.Y. (1988) Summary of the state-of-the-art in centrifuge model testing. In *Centrifuges in Soil Mechanics* (eds W.H. Craig, R.G. James and A.N. Schofield), pp. 11–18. Balkema, Rotterdam.

König, D., Jessberger, H.L., Bolton, M.D., Phillips, R., Bagge, G., Renzi, R. and Garnier, J. (1994) Pore pressure measurements during centrifuge model tests—experience of five laboratories. *Centrifuge '94,* pp. 101–108. Balkema, Rotterdam.

Kusakabe, O. (1982) Stability of Excavations in Soft Clay. Doctoral thesis, Cambridge University.

Kutter, B.L., Sathialingam, N. and Herrman, L.R. (1988) The effects of local arching and consolidation on pore pressure measurements in clay. *Centrifuge '88* (ed. J.F. Corté), pp. 115–118. Balkema, Rotterdam.

Lee, F.H. (1991) Frequency response of diaphragm pore pressure transducers in dynamic centrifuge model tests. *ASTM Geotech. Test. J.,* **13** (3), 201–207.

McVay, M., Bloomquist, D., Vanderlinde, D. and Clausen, J. (1994) Centrifuge modelling of laterally loaded pile groups in sand. *ASTM Geotech. Test. J.,* **17** (2), 129–137.

Miyake, M., Akamoto, H. and Aboshi, H. (1988) Filling and quiescent consolidation including sedimentation of dredged marine clays. *Centrifuge '88* (ed. J.F. Corté), pp. 163–170. Balkema, Rotterdam.

Nunez, I.L. and Randolph, M.F. (1984) Tension pile behaviour in clay—centrifuge modelling techniques. In *The Application of Centrifuge Modelling to Geotechnical Design* (ed. W.H. Craig), pp. 87–102. Balkema, Rotterdam.

Phillips, R. and Sekiguchi, H. (1991) Water Wave Trains in Drum Centrifuge. Cambridge University Engineering Department Technical Report CUED/D-SOILS/TR249.

Pokrovskii, G.I. and Fiodorov, I.S. (1953) *Centrifugal Modelling in Structures Designing.* Gosstroyizdat Publishers, Moscow.

Schofield, A.N. (1978) Use of centrifuge model testing to assess slope stability. *Canad. Geotech. J.,* **15**, 14–31.

Schofield, A.N. (1980) Cambridge geotechnical centrifuge operations. *Géotechnique,* **20**, 227–268.

Schofield, A.N. and Zeng, X. (1992) Design and performance of an Equivalent-Shear-Beam Container for Earthquake Centrifuge Modelling. Cambridge University Engineering Department Technical Report CUED/D-SOILS/TR245.

Tsuchida, T., Kobayashi, M. and Mizukami, J. (1991) Effect of ageing of marine clay and its duplication by high temperature consolidation. *Soils and Found.,* **31** (4), 133–147.

Wilson, J.M.R. (1988) A Theoretical and Experimental Investigation into the Dynamic Behaviour of Soils. Doctoral thesis, Cambridge University.

Xuedoon, W. (1988) Studies of the design of large scale centrifuge for geotechnical and structural tests. In *Centrifuges in Soil Mechanics* (eds W.H. Craig, R.G. James and A.N. Schofield), pp. 81–92. Balkema, Rotterdam.

Yamaguchi, H., Kimura, T. and Fujii, N. (1976) On the influence of progressive failure on the bearing capacity of shallow foundations in dense sand. *Soils and Found.,* **16** (4), 11–22.

Zelikson, A. (1969) Geotechnical models using the hydraulic gradient similarity method. *Géotechnique,* **19**, 495–508.

Zimmie, T.F., Mahmud, M.B. and De, A. (1993) Application of centrifuge modelling to contaminant migration in seabed waste disposal. *Proc. 4th Canadian Conf. Marine Geotechnical Engineering,* St Johns, Newfoundland, Vol. 2, pp. 611–624, C-CORE, Memorial University of Newfoundland.

4 Retaining walls and soil–structure interaction
W. POWRIE

4.1 Embedded retaining walls

4.1.1 General principles

An embedded wall uses the passive resistance of the soil in front of the wall below formation level to counter the overturning effect of the lateral stresses in the retained ground (Figure 4.1). The provision of props in front of the wall will reduce the depth of embedment required for stability, but if there are props at more than one level, the problem becomes statically indeterminate even if the wall is analysed when it is on the verge of collapse.

Until the 1960s, embedded retaining walls were typically formed of steel sheet piles, and were used almost exclusively in granular deposits. In these soils, it is generally accepted that the movement of an embedded wall required to mobilize the full passive pressure of the soil in front is very much greater than that required to achieve active conditions in the retained ground. Under working conditions, therefore, it might reasonably be assumed that the lateral effective stresses on the back of the wall had fallen to the active limit, but that the lateral effective stresses in front of the wall should be calculated with the

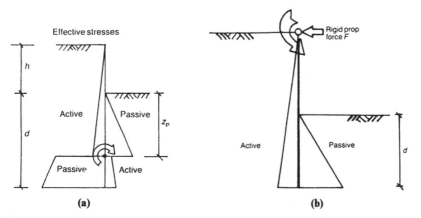

Figure 4.1 Idealized effective stress distributions at collapse for (a) an unpropped embedded wall and (b) an embedded wall propped at the crest. Reproduced from Bolton and Powrie (1987). Institution of Civil Engineers, with permission.

fully-passive earth pressure coefficient K_p divided by a factor F_p in the range 1.5 to 2. Equilibrium calculations carried out using a factored (reduced) value of K_p have traditionally formed the basis of design of embedded retaining walls in granular materials, the aim being that the wall should neither collapse outright nor deform excessively in service.

4.1.2 Embedded walls in dry granular material

Seminal research into the behaviour of embedded sheet pile retaining walls in sand was carried out during the 1950s by Rowe, who used both large scale laboratory tests at normal gravity ($1\,g$) and analytical methods. Rowe's $1\,g$ tests on unpropped walls (Rowe, 1951) showed that for this type of structure, conditions at collapse were reasonably well represented by the idealized theoretical distribution of effective stresses shown in Figure 4.1(a), with K_p based on the peak angle of shearing resistance ϕ'_{peak} and wall friction $\delta = 0.67\phi'_{peak}$. At factors of safety greater than unity, however, a triangular pressure distribution in front of the wall with the full passive pressure coefficient K_p reduced by a factor F_p was found to overestimate the lateral stresses near the toe of the wall, and hence bending moments. This is probably because with the wall rotating about a point close to the toe, there is only limited movement of the wall into the soil at this level, restricting the development of lateral stress.

Rowe's model tests on walls propped or anchored at the crest (Rowe, 1952) demonstrated that only a small movement of the prop or anchor was required for the earth pressures behind the wall to fall to the active limit. The effective stress distribution in front of a stiff wall was found to increase linearly with depth below formation level, consistent with the application of a reduction factor F_p to the full passive pressure coefficient K_p (Figure 4.1(b)). A wall whose deflexion at formation level is less than or equal to its deflexion at the toe may be defined for this purpose as stiff.

If the wall is more flexible, so that the deflexion at formation level is greater than that at the toe, the centroid of the lateral pressure distribution in the soil in front of the wall is raised. This leads to a reduction in bending moments and prop load to below the values obtained using the conventional limit equilibrium-based calculation. Rowe (1955) presented a design chart which can be used to apply a reduction factor to the calculated bending moments to allow for the redistribution of lateral stresses in the soil in front of a comparatively flexible retaining structure, such as a sheet pile wall. The main limitations of Rowe's moment reduction chart are that the reference calculation, in which it is assumed that fully active conditions are achieved in the soil behind the retaining wall, is applicable only to granular soils with initial *in situ* earth-pressure coefficients K_o close to the active limit, and that the relative wall flexibility is quantified in terms of an unusual and non-fundamental soil-stiffness parameter which is difficult to measure or even estimate reliably. It should also be noted that Rowe's experiments were carried out on dry sand: in a real situation with non-zero pore-water pressures, the moment reduction factor should only be applied to the proportion of the overall bending moment which is due to effective stresses.

Lyndon and Pearson (1984) reported the results of two centrifuge tests on rigid unpropped retaining structures, embedded in ballotini (small glass balls) of 150–250 μm diameter, with $\phi'_{crit} = 19.5°$ and $\phi'_{peak} = 38°$ (both measured in plane strain) at a density of $1600 \, \text{kg/m}^3$. Their retaining wall was 185 mm high (representing a real wall of overall length 11.1 m at a scale of 1:60) and 32 mm (1.92 m at field scale) thick. Both faces of the wall incorporated slots 12 mm wide $\times$ 57 mm long $\times$ 12 mm deep which, when filled with ballotini, would encourage the development of full friction over most of the soil–wall interface. Some of the slots housed boundary pressure cells comprising simply-supported strain-gauged beams, which were used to obtain stress distributions both behind and in front of the wall. Excavation in front of the wall was carried out in increments, stopping and restarting the centrifuge at each stage, until collapse occurred.

For problems of this type (i.e. retaining walls in dry sand), both the self-weight stresses which drive failure and the ability of the soil to resist shear increase with the applied g level. The results of the stress analyses shown in Figure 4.1 may be presented in terms of the embedment ratio d/h at failure as a function of the angle of shearing resistance ϕ' of the soil. The unit weight of the soil does not affect the embedment ratio at collapse, so that a wall which does not fail at 1 g should in theory be stable at any gravity level to which it is subjected in the centrifuge—provided that the angle of shearing resistance of the soil does not change. In reality, the peak angle of shearing resistance of a soil of a given void ratio (density) will decrease as the applied effective stress increases, due to the suppression of dilation. Thus a small-scale model may be stable at 1 g, but fail at a higher g level because the peak strength is reduced. If this happens, the implication is that a large retaining wall of a given embedment ratio will be less stable than a smaller wall of the same embedment ratio in the same material, because the peak angle of shearing resistance which maintains the stability of the smaller wall cannot be mobilized at the higher stresses which exist in the soil around the larger wall.

The implication of the foregoing is that considerable caution must be exercised in the selection of ϕ' values for the extrapolation of 1 g model test results to field scale structures. Rowe (1951, 1952) seems to have based his back-analyses on ϕ'_{peak}: in the application of his results to a larger structure, ϕ'_{peak} should be measured at the highest applicable stress level, in which case it may be similar to the critical state angle of shearing resistance, ϕ'_{crit}. A further point is that in small-scale laboratory tests at 1 g, the range of stress is not great and the use of a uniform value of ϕ'_{peak} in back-analysis is probably justifiable. At field scale, or in a centrifuge model, the range of stress is much larger, and (quite apart from the possibility of progressive failure) the existence of a uniform value of ϕ'_{peak} is extremely unlikely. In these circumstances, the use of ϕ'_{crit} rather than ϕ'_{peak} in back-analysis and design would seem to be appropriate.

Thus one reason for testing models of retaining walls in dry sand in a centrifuge rather than at $1\,g$ is that the stress state of the soil (and hence its stress–strain behaviour, including the potential or lack thereof for the development of peak strengths) in the centrifuge is similar to that for a large structure in the field. A second reason is that stresses, bending moments and prop forces are increased in proportion to the g level at which the model is tested, making them easier to measure reliably.

The two walls tested by Lyndon and Pearson (1984) would have been expected to behave identically, but they did not. The first wall deformed primarily by forward rotation about the toe, whereas the mode of displacement of the second wall was predominantly translation. Lyndon and Pearson suggest that this may have been due to the observed slight backward inclination of the second wall at the start of the test, which perhaps allowed the enhanced self-weight of this unusually thick retaining structure to act eccentrically, developing a restoring moment in the opposite sense to the overturning moment exerted by the retained soil. The effective stress distributions measured by Lyndon and Pearson are in both cases broadly consistent with those of Rowe (1951).

The first wall (which deformed by rigid body rotation about a point near the toe) failed catastrophically at an embedment ratio d/h of 0.414, but whether this was at the test acceleration of $60\,g$ or while the centrifuge acceleration was still increasing towards this value following the removal of the last layer of soil in front of the wall is not stated. The mobilized soil strength ϕ'_{mob} for an embedment ratio of 0.414 is approximately 36° according to the stress analysis shown in Figure 4.1(a), using the earth-pressure coefficients given by Caquot and Kerisel (1948) with wall friction $\delta = \phi'_{mob}$. Failure of the second wall was less readily identifiable, although at an embedment ratio of 0.512 its movement was quite large. The mobilized soil strength for an embedment ratio of 0.512 is just less than 32° (again assuming $\delta = \phi'_{mob}$). These values of ϕ'_{mob} at failure lie between the peak and critical state values quoted by Lyndon and Pearson, which does not seem unreasonable for a material (ballotini) with a rather stronger potential for dilation than most real soils. However, the difference between the embedment ratios at failure suggests that it would be unwise in a calculation forming the basis of a design to rely on the uniform mobilization of peak strengths at collapse. A further point which emerges from the second of the tests reported by Lyndon and Pearson is that the identification of failure is often not obvious: the comparison of data from different research workers may therefore be complicated by the various different definitions of failure they use.

King and McLoughlin (1993) summarized a series of tests on more flexible unpropped embedded cantilever walls, retaining dry sand of typical grain size

0.16 mm. The model wall was fabricated from stainless-steel sheet 2 mm thick, representing a Frodingham No. 5 section sheet pile wall of bending stiffness 10.2×10^4 kN m^2/m and total length 11 m, at a scale of 1:92. Tests were carried out on walls with three different surface finishes, giving angles of soil/wall friction $\delta = 0$, $\delta = \phi'_{peak}$ and intermediate values of $\delta = 0.36\phi'_{peak}$ to $\delta = 0.40\phi'_{peak}$, retaining either dense sand (density $\rho = 1631$ kg/m^3, $\phi'_{peak} = 49.5°$) or loose sand (density $\rho = 1448$ kg/m^3, $\phi'_{peak} = 40.0°$). Bending moments were measured directly by strain gauges glued to the wall, because it is not usually possible to incorporate boundary pressure cells into model embedded walls of realistic scale thickness. As with the tests reported by Lyndon and Pearson (1984), the tests of King and McLoughlin were carried out on the Liverpool University centrifuge, and the soil in front of the wall was excavated in stages by stopping and restarting the machine. This procedure is considered by King and McLoughlin not to have affected the behaviour of the model in any significant way: given the stabilizing effect of the likely increase in ϕ'_{peak} at centrifuge accelerations below the test value, this conclusion is probably not unreasonable.

Table 4.1 compares the embedment ratios (d/h) at collapse observed by King and McLoughlin with those predicted using the stress field calculation shown in Figure 4.1(a) with Caquot and Kerisel's earth-pressure coefficients for the values of ϕ' and δ stated. For the centrifuge tests, it is not possible to identify the embedment ratio at failure precisely: a range is given, corresponding to the last stable excavation depth, and the removal of a further 0.5 m of soil (at field scale) from in front of the wall, which resulted in collapse. It may be seen that the calculations using the peak angles of shearing resistance quoted by King and McLoughlin consistently overestimate the embedment ratio at failure, particularly in the case of the tests on dense sand. Unfortunately, the critical state angle of shearing resistance is not given by King and McLoughlin. Since, however, the predicted embedment ratios for the

Table 4.1 Comparison of actual and theoretical retained heights (in metres at field scale) at collapse for centrifuge model tests of unpropped walls in dry sand by King and McLoughlin (1992)

Sand	King and McLoughlin centrifuge test (m)	Figure 4.1(a) stress analysis (m)
Dense sand $\phi'_{peak} = 49.5°$		
$\delta = 0$	7.0–7.5	8.00
$\delta = 17.7°$	8.0–8.5	8.54
$\delta = 49.5°$	8.5–9.0	9.51
Loose sand $\phi'_{peak} = 40°$		
$\delta = 0$	6.0–6.5	6.89
$\delta = 15.8°$	7.0–7.5	7.46
$\delta = 40°$	8.0–8.5	8.36

loose sand models (for which ϕ'_{peak} is probably only slightly in excess of ϕ'_{crit}) are close to or just outside the upper limits observed in the centrifuge tests, it seems likely that the use of earth pressure coefficients based on ϕ'_{crit} in the stress field calculation shown in Figure 4.1(a) would lead to a generally correct or only slightly conservative prediction of the collapse limit state.

The use of a limit-type stress distribution with earth pressure coefficients based on peak soil strengths and a passive pressure reduction factor F_p was found by King and McLoughlin generally to overestimate maximum bending moments under conditions where $F_p = 1.5$. This is consistent with the earlier work of Rowe (1951), and arises because the centroid of the pressure distribution in front of the wall is higher than in the idealized distribution shown in Figure 4.1(a). For King and McLoughlin's rough walls with $\delta = \phi'_{peak}$, however, the maximum observed bending moments were very close to those calculated using the limit-type stress distribution. The measured bending moments in the rough walls were up to 80% higher than those in the smooth walls, and up to 50% higher than those in the walls of intermediate roughness. Wall roughness was not investigated by Rowe. Although there are insufficient data to be certain, and one possibility is that their calculations were based on unrealistically high values of ϕ' and therefore overestimated the real factors of safety quite significantly, King and McLoughlin's results would tend to militate against the use of moment reduction factors in design where the wall is very rough.

King and McLoughlin also observed that the deformations of all walls at $F_p = 1.5$ (based on peak strengths) were much larger than would be tolerated in reality. This demonstrates one of the major disadvantages of using a factor on passive pressure coefficient F_p in design, which is that the additional embedment required to increase the numerical value of F_p from 1.0 to 1.5 or even 2.0 may be very small, especially when the angle of shearing resistance of the soil ϕ' is high (Simpson, 1992). The problem is exacerbated by the inappropriate use of ϕ'_{peak} (rather than ϕ'_{crit}) as a design parameter.

4.1.3   Embedded walls in clay

Since the advent in the 1960s of *in situ* methods of wall installation, such as diaphragm walling and secant and contiguous piling, embedded retaining walls have become increasingly constructed from reinforced concrete to retain clay soils. It gradually became apparent that the use of factored limit-based methods of analysis in the conventional way led to the calculation of large depths of embedment (Hubbard *et al.*, 1984; Garrett and Barnes, 1984). This was primarily because the factor F_p was applied to the passive pressure coefficient rather than to the soil strength directly, and the numerical values of F_p which had been found satisfactory for use with granular soils having a critical state angle of shearing resistance ϕ'_{crit} of 30° or more could lead to apparently uneconomic depths of embedment in clay soils with ϕ'_{crit} values of only 20° or so. A further concern arose from the fact that the stress history of a typical overconsolidated clay deposit is such that the *in situ* lateral stresses are comparatively high, perhaps approaching the passive limit. Under these conditions, the conventional assumption that more movement is required to reach the active state than the passive is clearly suspect, and there is a possibility that a wall which moves sufficiently for the lateral stresses in the retained soil to fall to the active limit will have become unserviceable. Alternatively, if the embedment of the wall is sufficient to limit movement, the lateral effective stresses exerted by the retained soil might be considerably in excess of those calculated assuming fully active conditions. There was also (and there still is) some uncertainty concerning the length of time it would take for pore water pressures in clay soils to reach equilibrium following the construction and excavation in front of the wall, and—particularly for temporary works—the need to consider both short- and long-term conditions in design.

Bolton and Powrie (1987, 1988) carried out a series of centrifuge model tests using the Cambridge University centrifuge to investigate both the collapse and serviceability behaviour of diaphragm walls in clay. Figure 4.2 shows a cross-sectional view of a typical model, which represents a wall with a retained height of 10 m at a scale of 1:125, deforming in plane strain. Walls were made from either 9.5 mm or 4.7 mm aluminium plate, corresponding to full-scale flexural rigidities of approximately $10^7 \, \text{kN m}^2/\text{m}$ and $1.2 \times 10^6 \, \text{kN m}^2/\text{m}$, respectively.

The clay used in the model tests was speswhite kaolin, which was chosen primarily because of its relatively high permeability, $k = 0.8 \times 10^{-9} \, \text{m/s}$. The kaolin sample was consolidated from a slurry to a vertical effective stress of

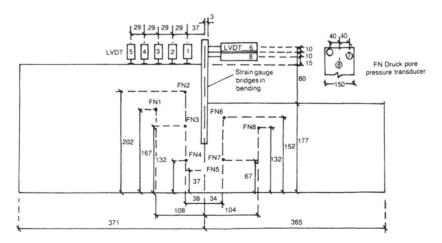

Figure 4.2 Cross-section through a model diaphragm wall (dimensions in millimetres at model scale). Reproduced from Bolton and Powrie (1987). Institution of Civil Engineers, with permission.

1250 kPa before being unloaded to a vertical effective stress of 80 kPa. The clay was removed from the consolidation press, and the model was prepared.

With tests on clay soils, pore-water pressures (and the control thereof) cannot realistically be neglected. The final stage in the preparation of a clay model before the test proper commences is therefore usually a period of reconsolidation in the centrifuge, during which the clay sample is brought into equilibrium under its enhanced self-weight at the appropriate g level. After this time, the clay is in a state which corresponds to idealized field conditions, with effective stresses increasing with depth and hydrostatic pore-water pressures below the groundwater level set by the modeller. The one possible exception to this is a mechanistic study, in which the g level might be increased steadily to initiate the rapid collapse of a model in a clay sample having a constant profile of undrained shear strength t_u with depth. In such a case, a low-permeability clay should be used to ensure that any changes in volume—and hence in undrained shear strength—which occur during the gradual increase in centrifuge acceleration are insignificant.

At the end of the reconsolidation period, it is necessary to simulate excavation in front of the wall without stopping the centrifuge. The most common technique, as used by Bolton and Powrie (1987, 1988), is to form the excavation before mounting the model on the centrifuge. The soil removed is replaced by a rubber bag filled with zinc chloride solution[1] mixed to the same unit weight as the clay. A valve-controlled waste-pipe is used to drain the zinc chloride solution from the rubber bag, simulating the excavation of soil from in front of the wall, at an appropriate stage following reconsolidation of the clay sample in the centrifuge.

The vertical stress history imposed on their clay samples by Bolton and Powrie corresponded to the removal by erosion of about 150 m of overlying soil, which is reasonably representative of a typical overconsolidated clay deposit. Although the *in situ* lateral stresses in such a soil are likely to be high, the slurry trench phase of diaphragm wall construction is certain to result in a significant alteration to this initial condition. The exact effect of the installation of the wall will depend on a number of factors, but an approximate analysis can be used to estimate the likely range of the pre-excavation lateral earth pressure coefficient (Powrie, 1985). In London clay, for example, the slurry trench phase might reduce an initial effective stress earth-pressure coefficient of 2.0 to between 1.0 and 1.2. A pre-excavation lateral earth-

---

[1] It should be noted that zinc chloride solution is highly corrosive to metals (especially aluminium), and irritant to skin. Extreme care should be taken in its handling and use. Protective clothing, including goggles and disposable rubber gloves, should be worn at all times. Splashes on the skin should be washed off immediately: splashes in the eyes should be irrigated with copious quantities of water. In cases of accidental swallowing, or severe external contact (especially with the eyes), seek medical advice. Zinc chloride solution drained from the model during the simulation of excavation should be retained for the remainder of the test in covered catch tanks: under no circumstances should it be vented into the centrifuge chamber. Plastic pipework is susceptible to attack by zinc chloride solution, and should be checked carefully after each test.

pressure coefficient K_o of unity might therefore be considered appropriate for centrifuge tests on model diaphragm walls, which start with the wall already in place. As the zinc chloride solution in the tests reported by Bolton and Powrie was mixed to the same unit weight as the soil it replaced, the lateral stresses in front of the wall above formation level were consistent with the condition $K_o = 1$ after reconsolidation in the centrifuge. The bending moments measured during reconsolidation were generally very small, indicating that $K_o = 1$ was quite closely achieved in the retained soil as well.

Although it might at some time in the future be possible to replicate exactly in a centrifuge the processes of diaphragm wall installation and excavation in front, this is at present an unattainable ideal. In investigating the long-term (post-excavation) behaviour of diaphragm walls in clay, it is therefore necessary to start the centrifuge test (like Bolton and Powrie) with the wall already in place. Even if it is attempted to replicate the likely pre-excavation pore pressures and effective stresses rather than their *in situ* values, the problem remains that in the centrifuge model the changes in stress between the *in situ* and pre-excavation conditions will probably have to be applied across the entire sample, whereas in reality they would be confined to the vicinity of the wall. It seems reasonable to argue, however, that the stress state of the soil remote from the wall should not influence the behaviour of the wall to any great extent, and that it is more important to model correctly the stress state of the soil adjacent to the wall.

A second concern relates to the mobilization of soil–wall friction during reconsolidation. The stress history of the clay samples used by Bolton and Powrie was such that overall soil surface settlements were observed during reconsolidation in the centrifuge. If there were any relative soil–wall movement during reconsolidation, the soil would have tended to move downward relative to the wall. On excavation, the relative soil–wall movement would have continued in the same sense as far as the retained soil was concerned, but its direction would have been reversed for the soil remaining in front of the wall. This might have increased the rate of mobilization of soil–wall friction on the active side of the wall (where it makes comparatively little difference to the earth pressure coefficients), but delayed the mobilization of soil–wall friction on the passive side (where its effect is considerable). Overall, therefore, the models would be expected to err on the conservative side in that they would tend to overestimate the displacements of a corresponding full-scale construction, in which the mobilized soil–wall friction at the start of the excavation process was zero.

The stress history of an overconsolidated clay deposit immediately prior to diaphragm wall installation might have been one-dimensional swelling (on geological unloading or an increase in groundwater level) or compression (following reloading or underdrainage). It might, in principle, be attempted to replicate any of these in carrying out a centrifuge model test. It should, however, be noted that the argument for conservatism from the point of view of the rates of mobilization of soil–wall friction would not necessarily apply if the stress history to which the sample had been subjected during preparation had been different.

The layout of the instrumentation installed in a typical model diaphragm wall as tested by Bolton and Powrie is shown in Figure 4.2. Bending moments were measured using strain gauges glued to the wall, which were water-proofed and protected by a coating of resin 2 mm thick.[2] The resin coating also served to increase the thickness of the model wall so that both the thickness and the flexural rigidity of a typical reinforced concrete field structure were modelled correctly. Pore-water pressures were measured using Druck PDCR81 miniature transducers, and soil-surface settlements were measured using Sangamo (Schlumberger) LVDTs. Black markers embedded into the front face of the model enabled vectors of soil movement to be measured from photographs taken at various stages during a test. The density of the kaolin used in the centrifuge tests was approximately 1768 kg/m^3, the critical state angle of shearing resistance $\phi'_{crit} = 22°$ and the peak angle of shearing resistance $\phi'_{peak} = 26°$. The angle of friction δ between the soil and the resin used to coat the model wall was found (in shear box tests carried out at appropriate normal stresses) to be the same as the critical state angle of shearing resistance of the soil.

In most tests, a full-height groundwater level on the retained side of the wall was modelled and purpose-made silicone rubber wiper seals were used to prevent water from leaking between the edges of the wall and the sides of the strongbox. During the initial reconsolidation period, water was supplied at the elevation of the retained ground to the soil surfaces behind and in front of the wall and to an internal drain formed by a layer of porous plastic overlying grooves machined into the strongbox at the base of the soil sample.[3] After excavation, solenoid valves were used to switch drainage lines to isolate the base drain and to keep the water level within the excavation drawn down to the excavated soil surface. It can be shown that the presence of the isolated drainage sheet at the base of the model results in a steady state seepage regime similar to that for a somewhat deeper clay stratum with an impermeable boundary at the base.

---

[2] Since most electrical resistance strain gauges have a similar gauge factor, the sensitivity (in terms of mV/kN m) of a Wheatstone bridge arrangement used to measure wall-bending moments will decrease as the flexural rigidity of the wall is increased. Strain gauges which are temperature-compensated for the material on which they are mounted should always be used, but even so, problems can still arise with stiff walls due to changes in the temperature of the leads and wiring, or the use of a thermally-unstable resin to coat the face of the wall. With an amplification factor of 100 applied to the signals from the bridges, the results from the strain gauge bridges on the walls made from 9.5 mm dural plate were just about acceptable for short-term readings in terms of noise and repeatability. It is therefore considered that the direct measurement of bending moments in this way is generally unlikely to be feasible with walls stiffer than this.

[3] The system of grooves was required because the porous plastic had only a limited permeability in lateral flow. The porous plastic can satisfactorily be replaced by a sand layer, separated from the clay by a sheet of filter paper.

In addition to the considerations of the stress state and the stress–strain behaviour of the soil, one of the main reasons for carrying out centrifuge model tests on *in situ* retaining walls in clays is that the time taken for the pore-water pressures to move towards their long-term equilibrium state following excavation in front of the wall are reduced by a factor of N^2 (in a $1/N$-scale model) compared with a full-size structure. Unless the wall is well supported by props during excavation, it is usually the long-term conditions which are the more critical. This is because excavation processes tend to cause transient pore pressures which are below the long-term steady-state values, aiding stability in the short-term. At a scale of 1:100, the changes in pore pressure which would occur over a period of approximately 14 months at field scale may be observed in one hour in a centrifuge model made from the same soil. A further reason for carrying out centrifuge rather than 1 g tests is that in a 1 g test, the relationship between the undrained shear strength and the self-weight stresses, which governs the short-term behaviour, is likely to be so unrepresentative of any real situation as to be effectively meaningless.

The centrifuge tests on model diaphragm walls carried out by Bolton and Powrie demonstrated quite graphically the calamitous effect that ground-water can have on these structures. With an excavated depth of 10 m at prototype scale and a full-height groundwater level in the retained soil, unpropped walls of 5 m and 10 m embedment failed almost immediately on excavation. The initial soil deformations were so large that a tension crack opened between the wall and the soil. Surface-water ponding in the settlement trough behind the wall filled the crack, pushing the wall over almost instantaneously. The retained ground was left standing in a cliff as shown in Figure 4.3. Although in a field situation surface water is unlikely to be as readily available as it was in the model tests, either a burst water main or a thin gravel aquifer could have a similar effect.

An analysis based on the limiting lateral stresses associated with the estimated undrained shear strength profile of the clay sample used in the centrifuge model tests indicates that the theoretical maximum depth of a tension crack is 5.7 m dry and 31 m flooded. In the absence of water to fill cracks, an unpropped wall retaining 10 m of clay would require only a small embedment of about 2 m to maintain short-term equilibrium. If a crack between the wall and the soil should flood, however, it could remain open to a considerable depth transferring hydraulic thrust to the wall. The wall would then be forced outwards as the crack widened, provided that the rate of inflow was sufficient to maintain the hydraulic head. According to a limit equilibrium stress analysis, the embedment at which a water-filled crack could no longer cause complete failure of the wall would be 14 m: this is consistent with the observed short-term behaviour of the model walls. With a nominally full-height groundwater level on the retained side, a wall of 15 m embedment suffered large movements on excavation but remained in contact with the soil. Wall movements continued after excavation with no sign of abatement. The

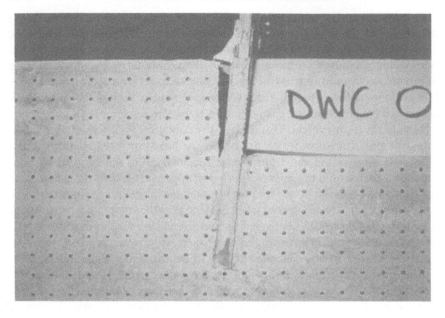

Figure 4.3 Flooded tension crack failure of unpropped model diaphragm wall in clay with $h = d = 10$ m at field scale. Reproduced from Bolton and Powrie (1987). Institution of Civil Engineers, with permission.

deformations were clearly influenced by the development, some time after excavation, of one or more slip surfaces: Figure 4.4 illustrates the final pattern of rupture lines. The deepest unpropped wall tested had an embedment of 20 m which, with a full-height groundwater level in the retained soil, was still insufficient to prevent unacceptably large ground movements.

The long-term behaviour of walls having sufficient embedment to prevent short-term collapse must be analysed in terms of effective stresses and pore water pressures. The equilibrium of the unpropped walls of 15 m and 20 m embedment at an instant near the end of each test was investigated using the assumed distribution of effective stresses shown in Figure 4.1(a), together with the measured pore water pressures. The soil strength ϕ'_{mob} which satisfied the equations of horizontal and moment equilibrium for the wall was found by iteration, assuming that the same soil strength applied on both sides of the wall and that the angle of soil–wall friction $\delta = \phi'_{mob}$. Earth-pressure coefficients were taken from Caquot and Kerisel (1948).

For the wall of 15 m embedment some time after excavation (corresponding to about six years at field scale), the back-analysis indicated a mobilized angle of shearing ϕ'_{mob} of 21.7°, almost identical to the critical state value. The analysis also predicted a pivot point at $z_p = 14$ m below formation level, which compares well with the measured soil displacements which indicated a

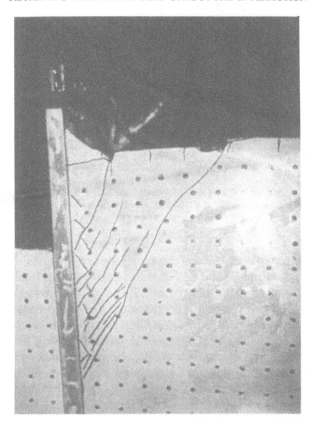

Figure 4.4 Rupture pattern in the clay retained by an unpropped model diaphragm wall with $h = 10$ m and $d = 15$ m at field scale. Reproduced from Bolton and Powrie (1987). Institution of Civil Engineers, with permission.

pivot at 13.5 m depth. The main rupture surface shown in Figure 4.4 intersected the wall at about 10 m below formation level. For the wall of 20 m embedment, the mobilized angle of shearing required for equilibrium in the long-term (i.e. after about 14 years at field scale) was 19.7°, slightly smaller than the critical state angle. This is consistent with the less-damaging deformation and the absence of any clear rupture surface, but the wall would in practice be deemed to have suffered a serviceability failure.

In the case of a wall of 5 m embedment propped at the crest, significant rupture surfaces developed in the soil at a time corresponding to approximately two years at field scale after excavation (Figure 4.5). The steepness of the main rupture surface near the top of the wall is due to the kinematic restraint imposed by the prop: pure sliding would be impossible at this level along a non-vertical slip surface. With the pore water pressures measured at

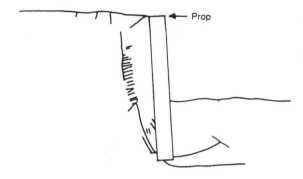

Figure 4.5 Rupture pattern in the vicinity of a model diaphragm wall propped at the crest with $h = 10$ m and $d = 5$ m at field scale. Reproduced from Bolton and Powrie (1987). Institution of Civil Engineers, with permission.

the end of the test (7.38 years at field scale following excavation), the static equilibrium of the zones of soil bounded by the rupture surfaces requires the mobilization of soil strengths of between 22° and 26°. This is reasonable, since these values represent, respectively, the estimated critical state and peak strengths of the kaolin used in the centrifuge tests. However, the equivalent earth pressure coefficients deduced from the thrusts arising from the back-analysis of the collapse mechanism are 0.23 on the retained side and 4.48 in front of the wall. According to Caquot and Kerisel, these earth-pressure coefficients correspond to $\phi' = 35°$ and $\delta = \phi'_{crit} = 22°$ behind the wall, and $\phi = 26°$ and $\delta = 22°$ in front. Strengths as large as these on the retained side would not generally be invoked on the basis of the data from laboratory tests on soil elements, and the discrepancy results from the use of an earth-pressure coefficient which is conservative in that it takes no account of the kinematic restraint imposed by the prop. Tests were also carried out on walls of 10 m and 15 m embedment propped at the crest. In these cases the embedment was sufficient to prevent failure, and soil movements were correspondingly smaller.

For unpropped walls, it was observed that significant soil movements occurred mainly in the zones defined approximately by lines drawn at 45° extending upward from the toe of the wall on both sides. This pattern of deformations might be represented for an unpropped embedded wall VW rotating about a point O near its toe by the idealized strain field shown in Figure 4.6. For stiff walls in which bending effects are small, the shear strain increment in each of the six deforming triangles is uniform and equal to twice the incremental wall rotation $\delta\theta$. The stress analysis shown in Figure 4.1(a) can be used for an unpropped wall of any geometry to calculate the uniform mobilized soil strength required for equilibrium ϕ'_{mob}. The remoteness of the wall from failure could then be quantified by its factor of safety based on soil strength, $\tan \phi'_{crit} / \tan \phi'_{mob}$. It is more rational to apply a factor of safety to

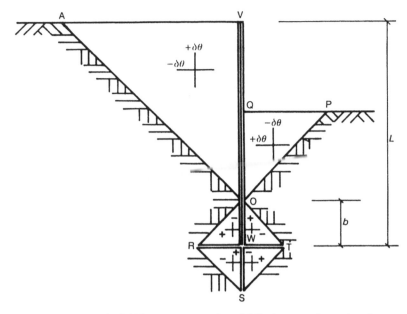

Figure 4.6 Idealized strain field for an unpropped model diaphragm wall rotating about a point near its toe. Reproduced from Bolton and Powrie (1988). Institution of Civil Engineers, with permission.

soil strength than to a parameter such as the depth of embedment or the passive pressure coefficient, which (as the centrifuge tests reported by King and McLoughlin show) can result in a wall which is actually very close to failure.

The shear strain corresponding to a given mobilized soil strength can be determined from an appropriate laboratory test on a representative soil element. The idealized strain field shown in Figure 4.6 can then be used to estimate the magnitude of the wall rotation and soil movements. For example, a calculation following Figure 4.1(a) based on the undeformed geometry and the pore-water pressures measured immediately after excavation in the centrifuge test on the unpropped wall of 20 m embedment indicates that a mobilized soil strength ϕ'_{mob} of 17.5° is required for equilibrium with $\delta = \phi'_{mob}$. The corresponding shear strain according to plane strain laboratory test data is 1.1%, which with the calculated pivot depth z_p of 18.8 m leads to a crest deflexion of 158 mm at field scale. Figure 4.7 shows that the measured and calculated soil movements are very similar. A similar calculation for walls propped at the crest is described by Bolton and Powrie (1988), and a worked example is given by Bolton et al. (1989, 1990a). The allowable mobilized soil strength for design purposes will depend on a number of factors, including the initial stress state of the soil and its stiffness measured following appropriate stress and strain paths.

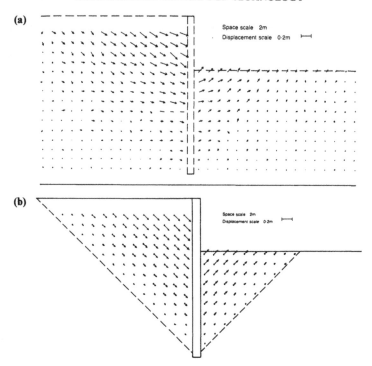

Figure 4.7 Comparison of (a) measured and (b) calculated soil movements during excavation in front of an unpropped model diaphragm wall with $h = 10$ m and $d = 20$ m. Reproduced from Bolton and Powrie (1988). Institution of Civil Engineers, with permission.

It would seem that the limit-based stress distribution shown in Figure 4.1(a) may be more readily adapted (by using factored soil strengths) to the serviceability analysis of cantilever retaining walls in clays than in granular materials (cf. Rowe, 1951; Lyndon and Pearson, 1984; King and McLoughlin, 1993). If this is so, it is probably because the clay remaining in front of the wall below formation level is brought almost to passive failure by the removal of the overlying soil, so that very little wall movement is required to mobilize high horizontal stresses, even near the pivot.

4.1.4   Other construction and excavation techniques

The draining of fluid from a rubber bag to simulate the stress changes which result from excavation is sometimes criticized on the grounds that it is only possible to model a pre-excavation earth pressure coefficient of unity. This is not so: Lade *et al.* (1981) used paraffin oil with a density of 7.65 kg/m³ to give a pre-excavation earth pressure coefficient of approximately $(1 - \sin \phi')$ in a granular material. At the other extreme, Powrie and Kantartzi (1993) describe the use of sodium chloride solution of density 1162 kg/m³, contained in a rubber bag filled to above the level of the soil surface, to impose a profile of K_o with depth similar to the *in situ* conditions in an overconsolidated clay deposit in the field (Figure 4.8). Following reconsolidation, the salt solution is drained to the level of the soil surface to simulate the stress changes caused by excavation of a diaphragm wall trench under bentonite slurry. This technique is being used in an investigation into the effects of diaphragm wall installation in overconsolidated clays, currently in progress at Queen Mary and Westfield College (University of London) using the geotechnical centrifuge operated jointly with City University.

A second, more valid, criticism of the 'draining fluid from a rubber bag' method of simulating excavation in the centrifuge is that the rate of reduction of lateral stress at any depth is approximately constant until the fluid level reaches that depth, which would not necessarily be the case in a real excavation. Other methods of simulating excavation in the centrifuge include the removal of rigid supports (Craig and Yildirim, 1976). Alternatively, the soil within the excavation could be contained within a flexible porous fabric bag, which could be winched clear at the appropriate stage of the centrifuge test using an electric motor. The latter method is proposed by Ko *et al.* (1982), but it is not clear whether it has actually been successful in practice. Further-

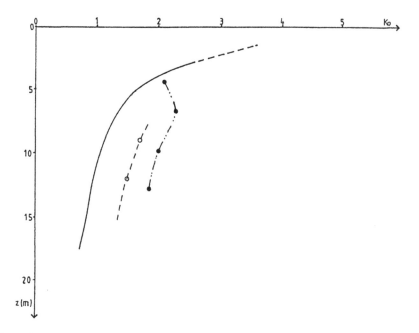

Figure 4.8 Pre-excavation effective earth pressure coefficients in centrifuge model tests to investigate installation effects of diaphragm walls in clay. ○ Tedd *et al.* (1984), cited in Powrie and Kantartzi (1992); ● Symons and Carder (1990)—centrifuge model. Reproduced from Powrie and Kantartzi (1993). Institution of Civil Engineers, with permission.

more, unless the excavation were to proceed in several stages using a number of fabric bags, the problem concerning the rate of lateral unloading would not be overcome.

The evolution of centrifuge modelling techniques to investigate the behaviour of embedded retaining walls has reflected the development of methods of construction of these structures in geotechnical engineering practice. Centrifuge model tests carried out by Powrie (1986) demonstrated the efficiency of props at formation level in terms of minimizing wall movements for a given embedment ratio. The main shortcoming of these tests (of which some of the results were summarized by Powrie and Li, 1991) was that the propping system had to be installed prior to reconsolidation and excavation in the centrifuge. New tests currently in progress at Queen Mary and Westfield College, in which the behaviour of *in situ* walls propped at both crest and formation level is under investigation, incorporate props with locking devices so that the props will not begin to take load or resist movement until they are required to do so by the modeller. Walls of this type are being increasingly used for underpasses at major interchanges on arterial roads in urban areas.

4.1.5   Centrifuge tests for specific structures

The model tests described so far were undertaken primarily to identify and investigate the fundamental mechanisms of deformation or collapse associated with a particular class of retaining structure. Centrifuge modelling is equally suited for use in connection with individual projects, especially where the effect of ground conditions is uncertain or some aspect of the proposed design is untried. An example of this is given by Rigden and Rowe (1975), who carried out centrifuge model tests to investigate the feasibility of using an unreinforced concrete diaphragm wall, circular on plan, to retain the ground around a proposed underground multi-storey car park in Amsterdam. One of the main issues was whether the hoop compression in the circular diaphragm wall would be sufficient to prevent tensile cracking due to longitudinal bending. The centrifuge model tests indicated that tensile stresses in the circular diaphragm wall could be substantially eliminated, provided that the depth of embedment below formation level was not excessive.

4.2   Gravity and L-shaped retaining walls

4.2.1   Background

The scientific study of gravity retaining walls can be traced back at least as far as the start of the 18th century, and was one of the subjects of Coulomb's celebrated memoir on statics (Heyman, 1972). Perhaps for this reason, the basic problem of the stability of gravity retaining structures does not seem to have been investigated very extensively using centrifuge modelling techniques. Centrifuge model studies on gravity-type retaining walls (including L-walls and bridge abutments) have tended to focus on the additional analytical difficulties which result from (for example) the application of a line load to the retained soil surface (Bolton and Mak, 1984), cyclic surcharge and lateral loading (Hird and Djerbib, 1993) and the effects of construction on a clay foundation (Bolton and Sun, 1991).

4.2.2 Line loads behind L-shaped walls

The model configuration used by Bolton and Mak (1984) to investigate the behaviour of an L-shaped retaining wall with a line load applied to the surface of the retained soil is shown in Figure 4.9. The dimensions shown are those of the model, which at a scale of 1:60 represented a field structure with a retained height of 8.82 m. The base and the rear face of the wall incorporated Cambridge contact stress transducers (Stroud, 1971), which was why the wall was so thick. The soil used in the experiments was dry 14/25 Leighton Buzzard sand, with $\phi'_{crit} = 33°$ and ϕ'_{peak} up to 48°, depending on the relative density. The stress conditions under which the values of ϕ'_{peak} were measured are not stated. At a centrifugal acceleration of 60 g, the load applied to the footing was increased until the retaining wall–footing system failed. A series of tests was carried out, in which the relative density of the sand (and hence the

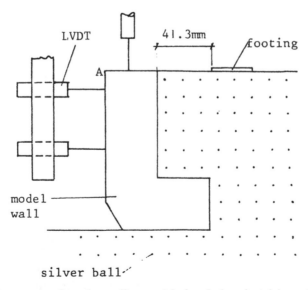

Figure 4.9 Cross-section through centrifuge model of an L-shaped retaining wall with a strip footing at the surface of the backfill. Total wall height $H = 147$ mm. Reproduced from Bolton and Mak (1984).

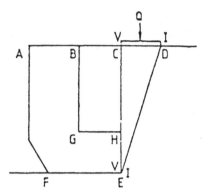

Figure 4.10 Idealized mechanism of failure for model wall–footing system. Reproduced from Bolton and Mak (1984).

peak angle of shearing resistance ϕ'_{peak}), the length B of the wall base, and the distance D_F between the footing and the retaining wall were varied.

From the displacement patterns and contours of shear strain measured during the centrifuge tests, an analysis based on the limiting equilibrium of a wedge of soil below the footing and a block comprising the wall and the soil contained within the 'L' was proposed (Figure 4.10). This analysis was used to relate the mobilized angles of shearing resistance ϕ'_{mob} on the planes VV and II to the load on the footing. Although values of ϕ'_{mob} on these planes in excess of ϕ'_{crit} were inferred, the rates of mobilization of soil strength and the instants of maximum mobilized soil strength along VV and II were not coincident: this would suggest that peak strengths are not mobilized uniformly at collapse. The centrifuge model tests indicated that the stability of the wall was controlled primarily by the bearing capacity of the base, which was in turn dependent on the soil strength mobilized along VV. The peak footing loads measured in the centrifuge, corrected for the effects of side friction in the model (which were in this case quite considerable), were generally over-predicted by the idealized wedge calculation. This is only to be expected from what is essentially an upper bound calculation, with the blocks ACEF and CDE forming a kinematically admissible mechanism analysed by statics rather than by a work balance.

4.2.3 Cyclic loading of L-shaped walls

Hird and Djerbib (1993) reported the results of centrifuge tests on stiff and flexible L-shaped cantilever walls, retaining dry 14/25 Leighton Buzzard sand, subjected separately to cyclic applications of lateral load to the crest of the wall (such as might be imposed by a fixed bridge deck due to temperature effects), and surcharge to the backfill. A cross-section through the model is shown (with dimensions at model scale) in Figure 4.11. At a scale of 1:60, the

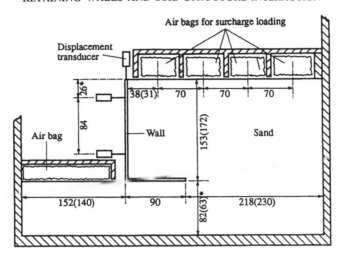

Figure 4.11 Cross-section through centrifuge model of an L-shaped retaining wall subjected to cyclic lateral loading at the crest and cyclic surcharge loading to the backfill. All dimensions in mm. Figures in parentheses refer to stiff wall. Asterisked dimensions increased by 19 mm in lateral loading tests. Reproduced from Hird and Djerbib (1993). Institution of Civil Engineers, with permission.

retained height of the field structure modelled is approximately 9 m. The more flexible wall was fabricated from 3.2 mm thick mild steel plate, giving a bending stiffness equivalent to a 0.4 m thick reinforced concrete wall stem, and was instrumented with strain gauges to measure bending moments. The stiff wall was made from 22 mm thick mild steel plate, representing a reinforced concrete structure approximately 3 m thick at field scale, and was instrumented with Stroud-type contact stress transducers. Hird and Djerbib consider that these models would represent the extremes of wall stiffness likely to be encountered in practice. The density of the sand as placed was 1700 kg/m³. The peak angle of shearing resistance ϕ'_{peak} was estimated on the basis of research by Stroud (1971) to be 46.5°, but the detailed reasoning behind this choice of ϕ'_{peak} is not given.

The cyclic surcharge was applied to the backfill by pressurizing each of four airbags behind the retaining wall in turn, simulating a 65 mm wide load (3.9 m at field scale) moving towards and then away from the wall. The lateral load was applied at the top of the wall by means of a double-acting pneumatic ram. A stiff beam was used to distribute the load evenly along the top of the wall during the pushing stroke, but the pulling force was transmitted to the wall by means of hooks at the edges, leading to some out-of-plane bending in the case of the more flexible wall. Certain other technical difficulties are described by Hird and Djerbib (1993).

The bending moments measured in the more flexible wall on reaching the required centrifuge acceleration of 60 g were consistent with active conditions based on the ϕ'_{peak} value of 46.5°. In the case of the stiff wall, the earth-pressure coefficients at this stage were (as might be expected) rather higher, increasing to just above K_{onc} with $\phi' = 46.5°$ ($K_{onc} = 1 - \sin \phi'$) at the base. In both cases, significant changes in bending moment/horizontal stress and lateral-wall movement were associated only with the application or removal of the surcharge closest to the wall. However, the residual bending moments/horizontal stresses and wall movements at zero surcharge gradually increased as cycling continued, with no steady-state value apparent after 15–20 load cycles.

The stiff wall showed more significant reductions in horizontal stress and larger movements back towards its initial position following the removal of each surcharge than the more flexible wall: in other words, the 'locking in' of horizontal stresses following the application of a surcharge was more complete for a more flexible retaining wall. This is qualitatively consistent with the idealized theories for lateral stresses induced by compaction described by Broms (1971) and Ingold (1979). These theories were developed for use with walls where the backfill is compacted in comparatively thin layers, however, and so are not strictly applicable in the present case. Some indication of the complexity of the problem addressed by Hird and Djerbib (1993) is given by the changes in the stress state of the soil indicated by the Stroud cells in the case of the stiff wall, which showed a reversal in the direction of the shear stress at the soil–wall interface during each load cycle. While this behaviour

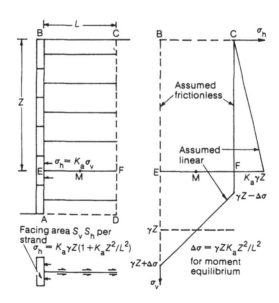

Figure 4.12 Stress analysis for reinforced soil retaining wall. $T_{max} = K_a \gamma Z(1 + K_a Z^2/L^2)S_v S_h$ (equation 4.3). Reproduced from Bolton and Pang (1982). Institution of Civil Engineers, with permission.

might be expected in theory, it is nonetheless instructive to see it demonstrated in practice.

The tests in which the retaining walls were subjected to cycles of lateral load showed similar progressive changes ('ratcheting') of displacements and bending moments/lateral stresses. Although the difference between successive cycles diminished as the number of cycles increased, a true steady state had not been reached for either the stiff wall or the more flexible wall after 50 cycles, when the centrifuge tests were terminated.

Where cyclic loads are expected to be applied throughout the life of a structure, their effects must be taken into consideration by the designer. While it is some reassurance that the rates of increase of deflexion and bending moment change are smaller at the high end of the cycle (which would normally be the more critical condition for design) than at the low end, appropriate design assumptions are generally far from clear. Although the failure of plane retaining walls following a progressive increase in lateral stresses due to the cyclic effects may be rare, there are certain classes of structure (for example reinforced concrete ring filter beds) for which it is certainly a problem (England, 1992). This is an area of considerable uncertainty at present, in which further research is required.

4.3 Reinforced soil walls, anchored earth and soil nailing

4.3.1 Reinforced soil walls

In addition to the modes of collapse relevant to conventional retaining walls (monolithic rotation or sliding, triggering of a landslide, and failure of the materials used to construct the wall itself), the designer of a reinforced soil wall must guard against the possibility of the failure of the reinforcement, either in tension or by slippage. Bolton and Pang (1982) presented a simple yet rational stress analysis (Figure 4.12), based on the limiting equilibrium of a block of soil BCFE behind the retaining wall. This block of soil has width L (the length of the reinforcement strips) and a general depth Z. It is assumed that the vertical boundaries BE and CF are frictionless, and that the horizontal effective stresses σ'_h on CF increase linearly with depth Z, with $\sigma'_h = K_a \sigma'_v$ and $\sigma'_v = \gamma Z$ in the absence of pore-water pressures, where γ is the unit weight of the retained soil, K_a is the active earth-pressure coefficient $K_a = (1 - \sin \phi')/(1 + \sin \phi')$ and ϕ' is the angle of shearing resistance of the soil. The vertical stress distribution along the base EF of the block of soil is assumed to decrease linearly from the back of the facing panels to the end of the reinforcement, with a mean value of γZ. The maximum value of the vertical stress on this boundary is obtained from the condition of moment equilibrium about M, the midpoint of EF. This analysis may be used to estimate the load on each facing panel and reinforcement strip on the verge of failure, and hence the factors of safety against tensile failure, F_T, and frictional failure (slippage) F_F, in the most critical reinforcement strip:

$$F_T = P/[K_a \gamma Z S_v S_h (1 + K_a Z^2/L^2)] \qquad (4.1)$$

$$F_F = 2BL\mu/[S_v S_h K_a (1 + K_a Z^2/L^2)] \qquad (4.2)$$

where P is the reinforcement load at tensile failure, S_v and S_h are the height and width of a facing panel, respectively, B is the width of a reinforcement strip, and μ is the coefficient of friction between reinforcement strip and the soil. The load on a reinforcement strip, T_{max}, at depth Z is $K_a \gamma Z S_v S_h (1 + K_a Z^2/L^2)$ (equation 4.3, Figure 4.12), and its pull-out resistance is $2BL\mu K_a \gamma Z$.

Prior to Bolton and Pang (1982), attempts to investigate the validity of this approach had centred on model tests carried out at $1\,g$, on reinforced soil retaining walls 200–500 mm in height. Generally, the stress analysis shown in Figure 4.12 had been found to underestimate the observed tensile strength of the models by factors of up to 1.8. Difficulties were also encountered with local weaknesses at the joints between the facing panels and the reinforcement strips. The main problems with $1\,g$ model tests on reinforced soil retaining walls in granular soils relate to (i) the low stress levels in the model, which permit the development of peak strengths ϕ'_{peak} almost uniformly down the back of the wall that would be unlikely to occur in the field, (ii) the delicate reinforcement needed because the soil stresses are low, which means that considerable additional stiffness will result from the attachment of strain gauges, and (iii) that significant local imperfections (for example at the joints with the facing panels) are likely to occur. These problems can be overcome by testing models at an enhanced gravity level in the geotechnical centrifuge.

Bolton and Pang (1982) report the results of a series of plane strain centrifuge tests on model reinforced soil walls, 200 mm high, retaining dry sand of density 1723 kg/m³. The peak angle of shearing resistance of the sand ϕ'_{peak} was 49° at a vertical effective stress of 50 kPa, and 43° at an effective stress of 400 kPa. The facing units were flexible, and four different types of reinforcement strip were used, with friction coefficients $\mu = \tan \delta$ between 0.16 and 0.75. The ratio of the reinforcement length to the retained height L/H varied between 0.5 and 1.5. In each case, the centrifuge acceleration was gradually increased until collapse occurred. Tensile failure of reinforced soil retaining walls may be investigated in this way because the factor of safety F_T (equation 4.1) will reduce as the unit weight of the soil in the centrifuge acceleration field $\gamma (= \rho n g)$ is increased with increasing centrifuge acceleration ng. Although the factor of safety against pull-out (slippage) of the reinforcement F_F shows no direct dependence on the unit weight in the centrifuge acceleration field, the suppression of dilation as the self-weight stresses of the soil are increased will in actuality have some effect because of the resulting reduction in peak soil strength ϕ'_{peak}.

Collapse occurred as a result of the failure of the reinforcement in tension in

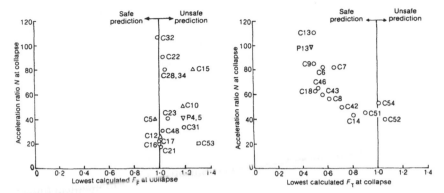

Figure 4.13 Values of (a) F_F and (b) F_T at collapse of centrifuge models of reinforced soil retaining walls. (a) Mark I tests (C): ○ SS (stainless steel), $L/H = 1.00$ (C31) to 0.43 (C48), △ MS (mild steel), $L/H = 0.75$; □ AL (aluminium), $L/H = 0.50$. Mark II tests (P) ▽ SS, $L/H = 0.80$. (b) Mark I tests (C): ○ AL, $L/H = 1.5$ (C6, 7, 8) to 0.5 (C14, 18, 51, 52, 54). Mark II tests (P): ▽ AL, $L/H = 0.5$. Reproduced from Bolton and Pang (1982). Institution of Civil Engineers, with permission.

some of the models, and by slippage of the reinforcement in others. The spread of values of F_T and F_F at collapse, calculated according to the stress analysis shown in Figure 4.12 and equations (4.1) and (4.2) respectively, is indicated in Figure 4.13. It may be seen that pull-out failure is generally well-predicted using equation (4.2), although there is a tendency to err on the unsafe side, with collapse in one test occurring at $F_F = 1.33$. According to Bolton and Pang (1982), this lack of conservatism could be reduced by the use of the minimum (rather than average) values of soil strength ϕ'_{peak} and soil–reinforcement coefficient of friction μ obtained from laboratory tests, to give factors of safety F_F at collapse of unity $\pm$ 15%, which is within the range of experimental error and repeatability. The use of critical state soil strengths might well have eliminated entirely the tendency of the calculation sometimes to err on the unsafe side, and is perhaps therefore to be recommended in design.

In contrast, the values of F_T at collapse (calculated according to equation 4.1) vary between 0.5 and 1.06, indicating that the stress analysis shown in Figure 4.12 is generally (but not always) unduly conservative. Tensile failure was well predicted by equation (4.1) only in tests on walls with narrow reinforced zones ($L/H = 0.5$), which were already close to pull-out failure. In other tests, it seemed that some redistribution of reinforcement loads was possible after the most critical reinforcement strip had yielded, thereby delaying the onset of collapse.

In order to clarify the apparent overconservatism of the stress analysis shown in Figure 4.12 in predicting tensile failure, Bolton and Pang carried out a second series of centrifuge tests. The basic geometry of the model was the same as that of the first series, but the reinforced soil zone was underlain by a wooden, rather than a sand, foundation. This enabled the vertical stress distribution at the base of the reinforced zone (AD in Figure 4.12) to be measured, without significantly affecting the behaviour of the model (some of the series 1 tests were duplicated with the new arrangement to verify this). Wall facings were either flexible (0.15 mm thick aluminium foil, as before) or more rigid (1 mm thick aluminium plate), and strain gauges were fixed to certain of the reinforcement strips in order to measure the tensile loads developed. The breaking strain of the 4 mm × 0.1 mm aluminium reinforcement strips was 0.35%, which was considered to be representative of a typical field situation in which extensive ductility of the reinforcement could not be relied on due to localized corrosion and/or construction defects.

In the second series of tests, the distribution of maximum reinforcement load with depth was found to be similar for five nominally identical models (having $L/H = 0.8$), up to a centrifuge acceleration of 63 g (Figure 4.14). It may be seen that the stress distribution shown in Figure 4.12 leads to the overprediction of the tension in the layer of reinforcement at the base of the wall by a factor of about 1.4, and that the most critical level (at which the tensions are greatest) is at a depth (below the soil surface) of 150 mm or $0.75H$. This is probably the result of the reduction in vertical stresses in this zone (as indicated by the pressure cells incorporated into the foundation), due to shear stresses between the facing panel and the soil, and the frictional resistance of the foundation which tends to limit the lateral movement of the lowest facing panel. This is one of the reasons why equation (4.1) is apparently overconservative in its prediction of tensile failure. However, the vertical stress reduction

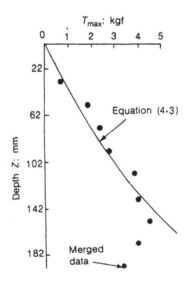

Figure 4.14 Peak reinforcement load as a function of depth for centrifuge models of reinforced soil retaining walls at 63 g. For equation (4.3) see Figure 4.12. Reproduced from Bolton and Pang (1982). Institution of Civil Engineers, with permission.

at the base of the wall might not occur if the facing panel were smooth, or if the foundation layer were comparatively compressible.

The centrifuge acceleration at collapse of Bolton and Pang's five series 2 models with $L/H = 0.8$ was in the range 63–$92\,g$. In the model which collapsed at $63\,g$, one reinforcement strip had failed at a reduced load near to its joint with the facing panel at $30\,g$, due to fatigue resulting from its repeated use in successive models. This precluded the possibility of the redistribution of reinforcement loads, leading to instant collapse following the yield of the reinforcement at the critical location. In another model which failed at $70\,g$, two reinforcement strips had ruptured at or soon after $63\,g$. In the model which collapsed at $92\,g$, significant plastic redistribution of reinforcement loads away from the critical zone had occurred, increasing the load carrying capacity of the construction as a whole by a factor of nearly 1.5. Plastic redistribution of reinforcement loads is the second reason for the apparent overconservatism of equation (4.1). This cannot be relied on, however, if the reinforcement is effectively brittle (for example, due to local corrosion or a weak joint with the facing panel), or if the factor of safety against pull-out F_F is close to unity.

Further tests showed that walls with a higher factor of safety against pull-out F_F are likely to be subjected to higher lateral earth pressures under working conditions, with $\sigma'_h = K_{onc}\sigma'_v$ (where $K_{onc} = 1 - \sin\phi'$) rather than $\sigma'_h = K_a\sigma'_v$ (Figure 4.15). Earth pressures at collapse, however, are slightly overpredicted (due to the effects of friction at the base of the wall and plastic stress redistribution) by the use of the active earth pressure. (In back-analyses,

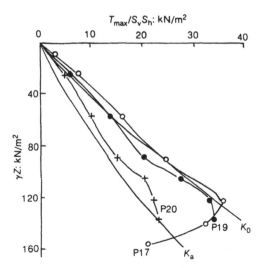

Figure 4.15 Earth pressures mobilized at $50\,g$ in models remote from tensile failure: frictional capacity of models P17 and P19 was double that of P20. Reproduced from Bolton and Pang (1982). Institution of Civil Engineers, with permission.

Bolton and Pang used ϕ'_{peak} values measured in shear box tests at a vertical effective stress equal to that at the base of the model wall at the centrifuge acceleration in question.)

The comprehensive series of centrifuge model tests described by Bolton and Pang gives a considerable insight into the behaviour of reinforced soil retaining walls in granular materials. In particular, the collapse of these walls can be conservatively predicted using the simple stress analysis shown in Figure 4.12, based on the peak angle of shearing resistance of the soil, measured at the appropriate vertical effective stress. If data on appropriate values of ϕ'_{peak} are not available, the use of ϕ'_{crit} would be acceptable in design. The apparent overconservatism which occurs in certain cases is due to factors which may be difficult to quantify or even control, including the exact details of wall construction, and cannot necessarily be relied upon. Furthermore, a degree of conservatism in design is necessary because it was not generally possible in the centrifuge tests to identify incipient collapse from an increase in the rate of wall movement with increasing g level or decreasing factor of safety. Empirical modifications to the stress analysis of Figure 4.12, involving for example the assumption of an erroneous 'mechanism' of failure, might not be appropriate in every case, and should therefore be used with considerable caution.

Mitchell et al. (1988) describe a series of centrifuge model tests on reinforced soil retaining walls, in which the effects of reinforcement type, facing panel stiffness, foundation stiffness and surcharge loading were investigated. The model walls were 150 mm high, and (as with Bolton and Pang's tests) were subjected to an increasing centrifuge acceleration until collapse occurred. The reinforcement strips were not strain-gauged, but the values of F_T at collapse, and data from earth-pressure cells, were generally supportive of Bolton and Pang's observations and hypotheses. Walls with stiffer facings collapsed at generally higher g levels than similar walls with flexible facings. Walls with aluminium foil reinforcement strips tended to fail catastrophically, whereas walls with plastic or geotextile reinforcement strips were more ductile.

The vertical stresses measured near the base of the wall were generally in excess of the overburden pressure, which is consistent with the trapezoidal distribution of base pressure indicated in Figure 4.12. With stiffer facing panels, however, the vertical stresses in this region were reduced, which is consistent with Bolton and Pang's observations and the hypothesis that stiff facing panels carry some of the vertical load in shear. Mitchell et al. tested some walls on compressible foundations of foam rubber or compacted clay. Not surprisingly, this did not affect significantly the factor of safety F_T at outright collapse for walls with flexible facings. Deformations prior to collapse were increased significantly, especially where both the facing panels and the reinforcement strips were of low stiffness. Unfortunately, no walls with stiff facings were tested on the compressible foundation: the settlement of a compressible foundation would be expected to destroy the capacity of friction at a stiff facing to reduce vertical stresses near the base of the wall, so that an increase in factor of safety (manifest by a reduction in g level) at collapse towards the value for a wall with flexible facings should be anticipated in such a case.

Mitchell *et al.* also report a limited number of tests on reinforced walls with a surcharge load at the retained soil surface. The data are too few to draw general conclusions; however it seems that (as with the walls with no surcharge) failure by pull-out tends to occur at $F_F \simeq 1$, while the prediction of tensile failure remains generally somewhat conservative. Two problems with ductile reinforcements are demonstrated. The first is that gross deformations occur without pull-out or breakage of the reinforcement, which implies that a wall which is to meet the usual requirements of serviceability must be designed to be very remote from either of these collapse limit states. The second is that certain reinforcements may creep, leading to failure after a certain time at a factor of safety greater than that which promotes instantaneous collapse. This is demonstrated by two tests carried out by Mitchell *et al.*, in which identical model walls with fabric sheet reinforcement collapsed instantaneously at $20\,g$, and after 70 min at $13\,g$. The problem of creep of geotextiles is well known and well researched in its own right, and must be addressed by centrifuge modellers investigating the behaviour of structures incorporating these materials, especially if the 'increasing g level to failure' method of testing is adopted.

4.3.2   Soil nailing

In recent years, soil nailing has developed as an economical method of retaining (at least temporarily) near-vertical cuts in clays and other soils. The soil nails are essentially reinforcing bars installed into the soil through the cut face, attached to a wire mesh-reinforced shotcrete covering. The nails may be installed in pre-drilled holes and grouted in place, or they may be fired into the ground using a compressed air gun.

Shen *et al.* (1982) carried out five centrifuge model tests to investigate failure mechanisms of nailed soil support systems in a silty sand. The retained height of the model was 15 cm in all cases, but the length and spacing of the grouted piano wire reinforcing elements were varied from test to test. Deformations were monitored using photography and video recordings, but nail loads were not measured. The method of testing was to increase the centrifuge acceleration until failure occurred. Failure was defined by Shen *et al.* as the onset of surface cracking, because the models consistently suffered large deformations without sudden collapse.

The results of the model tests were compared with analyses based on the limiting equilibrium of a block of soil defined by a parabolic arc through the base of the cut face. The measured and calculated g levels at failure were generally close (i.e. failure tended to occur at a factor of safety of unity), and the cracks observed in the retained soil surface were reasonably consistent with the positions of the critical parabolae. As the parabola does not constitute a kinematically admissible slip surface, the analysis is of the 'limit equilibrium' type, rather than a true upper bound. From an analytical point of view, there would seem to be little difference between a reinforced soil retaining wall and a nailed soil support system. The universal applicability of the method proposed by Shen *et al.* is perhaps, therefore, open to question— especially in situations (such as a compressible foundation layer) where the factors which lead to the general overconservatism of the stress analysis shown in Figure 4.12 are no longer present. However, it might be argued that a less conservative analysis is justified if the wire mesh-reinforced shotcrete facing is comparatively stiff (and therefore able to reduce vertical stresses at the base of the retained soil by friction), and because some of the vagaries of reinforced soil systems associated with tolerance gaps between the facings and uncertain construction practices are eliminated.

4.4 Concluding remarks

This chapter has described some examples of the ways in which geotechnical centrifuge models have been used to enhance our understanding of the behaviour of earth retaining structures. In some cases, a comparatively limited number of tests has served to corroborate previous work, and perhaps revealed some limitations (Lyndon and Pearson, 1984, and King and McLoughlin, 1993, on embedded cantilever walls in dry granular materials). In other cases, the behaviour of a particular class of retaining structure has been investigated comprehensively for the first time (Bolton and Powrie, 1987, 1988, on diaphragm walls in clay; Bolton and Pang, 1982, on reinforced earth retaining walls). Sometimes, new areas of uncertainty are revealed, which may have very significant implications in design (Hird and Djerbib, 1993, on cyclic lateral loading of L-shaped walls). In only a few reported cases has the centrifuge been used directly as part of an individual design (Rigden and Rowe, 1975, on an unreinforced circular diaphragm wall).

Centrifuge model tests have not only led to the clarification of mechanisms of deformation and collapse, but have also demonstrated that the use of the critical state soil strength will generally lead to the most reliable (if sometimes conservative) calculation of conditions at collapse. The evidence suggests quite strongly that it is unreasonable to assume that peak values of ϕ' will be mobilized uniformly throughout the soil mass at failure. If non-critical state strengths are used, for example in extrapolating the behaviour of small-scale models tested at $1\,g$ to field structures, the values of ϕ' should be selected very carefully with reference to the relative densities and the maximum effective stresses which apply to each case.

The constraints of centrifuge modelling can be significant, and their effects on the behaviour of the model must be considered in the interpretation of centrifuge model test results. A particular example of this is the way in which excavation in front of an *in situ* or sheet pile retaining wall is simulated. Nonetheless, it is doubtful that any of the work described in this chapter could have been carried out to quite the same effect without using a geotechnical centrifuge. In granular materials, it is the realistic variation of self-weight stresses with depth which forces the geotechnical engineer to address the crucial question of the appropriate value of ϕ' to use in calculations. In clay soils, centrifuge modelling also enables the effects of excess pore pressure dissipation to be observed under a reasonable timespan. The potential of the geotechnical centrifuge has been amply demonstrated over the last two decades by the many successful general mechanistic studies which it has been used to carry out. It is a source of regret that this potential has not yet been fully realized in the application of centrifuge modelling techniques to individual problems in civil engineering design.

References

Bolton, M.D. and Mak, K. (1984) The application of centrifuge models in the study of an interaction problem. *Proc. Symp. Application of Centrifuge Modelling to Geotechnical Design, Manchester, 16–18 April 1984* (ed. W.H. Craig), pp. 403–422. Balkema, Rotterdam.

Bolton, M.D. and Pang, P.L.R. (1982) Collapse limit states of reinforced earth retaining walls. *Géotechnique*, **32** (4), 349–367.

Bolton, M.D. and Powrie, W. (1987) Collapse of diaphragm walls retaining clay. *Géotechnique*, **37** (3), 335–353.

Bolton, M.D. and Powrie, W. (1988) Behaviour of diaphragm walls in clay prior to collapse. *Géotechnique*, **38** (2), 167–189.

Bolton, M.D. Powrie, W. and Symons, I.F. (1989) The design of stiff *in situ* walls retaining overconsolidated clay. Part I. Short term behaviour. *Ground Eng.*, **22** (8), 44–48.

Bolton, M.D., Powrie, W. and Symons, I.F. (1990a) The design of stiff *in situ* walls retaining overconsolidated clay. Part I. Short term behaviour. *Ground Eng.*, **22** (9), 34–40.

Bolton, M.D., Powrie, W. and Symons, I.F. (1990b) The design of stiff *in situ* walls retaining overconsolidated clay. Part II. Long term behaviour. *Ground Eng.*, **23** (2), 22–28.

Bolton, M.D. and Sun, H.W. (1991) The displacement of bridge abutments on clay. *Centrifuge '91* (eds H.-Y. Ko and F.G. McLean), pp. 91–98. Balkema, Rotterdam.

Broms, B.B. (1971) Lateral earth pressures due to compaction of cohesionless soils. *Proceedings of the 4th Budapest Conference on Soil Mechanics and Foundation Engineering (3rd Danube-European Conference)* (ed. A. Kezdi), pp. 373–384. Akademiai Kiado, Budapest.

Caquot, A. and Kerisel, J. (1948) *Tables for the calculation of passive pressure, active pressure and bearing capacity of foundations.* Gauthier Villars, Paris.

Craig, W.H. and Yildirim, S. (1976) Modelling excavations and excavation processes. *Proc. 6th Eur. Conf. Soil Mech. Found. Eng.*, Vol. 1, pp. 33–36.

England, G.L. (1992) Ring-Tension Filter Beds Subjected to Temperature Changes. MACE Centre Short Course Notes: The Behaviour of Granular Materials and their Containment. Imperial College, London.

Garrett, C. and Barnes, S.J. (1984) The design and performance of the Dunton Green retaining wall. *Géotechnique*, **34** (4), 533–548.

Heyman, J. (1972) *Coulomb's Memoir on Statics: An Essay in the History of Civil Engineering.* Cambridge University Press.

Hird, C.C. and Djerbib, Y. (1993) Centrifugal model tests of cyclic loading on L-shaped walls retaining sand. *Retaining Structures* (ed. C.R.I. Clayton), pp. 689–701. Thomas Telford, London.

Hubbard, H.W., Potts, D.M., Miller, D. and Burland, J.B. (1984) Design of the retaining walls for the M25 cut and cover tunnel at Bell Common. *Géotechnique*, **34** (4), 495–512.

Ingold, T.S. (1979) The effects of compaction on retaining walls. *Géotechnique*, **29** (3), 265–283.

King, G.J.W. and McLoughlin, J.P. (1993) Centrifuge model studies of a cantilever retaining wall in sand. *Retaining Structures* (ed. C.R.I. Clayton), pp. 711–720. Thomas Telford, London.

Ko, H.-Y., Azevedo, R. and Sture, S. (1982) Numerical and centrifugal modelling of excavations in sand. *Deformation and Failure of Granular Materials* (eds P.A. Vermeer and H.J. Luger), pp. 609–614. Balkema, Rotterdam.

Lade, P.V., Jessberger, H.L., Makowski, E. and Jordan, P. (1981) Modelling of deep shafts on centrifuge tests. *Proc. 10th Int. Conf. Soil Mech. Found. Eng.*, Vol. 1, pp. 683–691.

Lyndon, A. and Pearson, R. (1984) Pressure distribution on a rigid retaining wall in cohesionless material. *Proc. Symp. Application of Centrifuge Modelling to Geotechnical Design, Manchester, 16–18 April 1984* (ed. W.H. Craig), pp. 271–281. Balkema, Rotterdam.

Mitchell, J.K., Jaber, M., Shen, C.K. and Hua, Z.K. (1988) Behaviour of reinforced soil walls in centrifuge model tests. *Centrifuge '88*, pp. 259–271. Balkema, Rotterdam.

Powrie, W. (1985) Discussion on 5th Géotechnique Symposium-in-Print. The performance of propped and cantilevered rigid walls. *Géotechnique*, **35** (4), 546–548.

Powrie, W. (1986) The Behaviour of Diaphragm Walls in Clay. PhD Thesis, University of Cambridge.

Powrie, W. and Kantartzi, C. (1993) Installation effects of diaphragm walls in clay. *Retaining Structures* (ed. C.R.I. Clayton), pp. 37–45. Thomas Telford, London.

Powrie, W. and Li, E.S.F. (1991) Finite element analysis of an *in situ* wall propped at formation level. *Géotechnique*, **41** (4), 499–514.

Rigden, W.J. and Rowe, P.W. (1975) Model performance of an unreinforced diaphragm wall. *Diaphragm Walls and Anchorages*, pp. 63–67. ICE, London.

Rowe, P.W. (1951) Cantilever sheet piling in cohesionless soil. *Engineering*, **51**, 316–319.

Rowe, P.W. (1952) Anchored sheet pile walls. *Proc. ICE*, **1** (1), 27–70.

Rowe, P.W. (1955) A theoretical and experimental analysis of sheet pile walls. *Proc. ICE*, **1** (4), 32–69.

Shen, C.K., Kim, Y.S., Bang, S. and Mitchell, J.F. (1982) Centrifuge modelling of lateral earth support. *Proc. ASCE, J. Geol. Engg. Div.*, **108** (GT9), 1150–1164.

Simpson, B. (1992) Retaining structures: displacement and design (32nd Rankine Lecture). *Géotechnique*, **42** (4), 541–576.

Stroud, M.A. (1971) The Behaviour of Sand at Low Stress Levels in the Simple Shear Apparatus. PhD Thesis, University of Cambridge.

Symons, I.F. and Carder, D.R. (1990) Long term behaviour of embedded retaining walls in overconsolidated clay. In *Geotechnical Instrumentation in Practice*, pp. 289–307. Thomas Telford, London.

Tedd, P., Chard, B.M., Charles, J.A. and Symons, I.F. (1984) Behaviour of a propped embedded retaining wall in stiff clay at Bell Common Tunnel. *Géotechnique*, **34** (4), 513–532.

5 Buried structures and underground excavations

R.N. TAYLOR

5.1 Introduction

The scope of the chapter is to consider the modelling of buried pipes and culverts, and trench, shaft and tunnel excavations in soils ranging from sands, where the soil behaviour is fully drained, to clays, where the soil behaviour is initially undrained followed by time-dependent diffusion effects as excess pore pressures dissipate. In practice, all are problems which are affected by the construction process and any modelling needs to be undertaken with care to ensure proper representation of an aspect of the structure or excavation being investigated. Installing buried pipes requires a trench to be excavated, the pipe placed and the trench backfilled. Any excavation process is difficult to replicate in a centrifuge model since the soil appears very heavy during centrifuge flight and any tools or equipment used for excavation need to be both small and lightweight. Also, laying a pipe during centrifuge flight and placing and compacting backfill in a controlled, staged operation is far from easy. Similarly, in a prototype tunnelling operation, a cavity is advanced and a lining placed behind as an incremental process which is virtually impossible to mimic in detail in a small centrifuge model. In addition, if time effects or seepage into the tunnel are to be investigated, realistic boundary conditions need to be imposed.

All these practical considerations pose enormous difficulties to the experimenter and a criticism often raised is that since construction effects are not reproducible, little of value can emerge from model studies. However, such criticism can be directed at any modelling study including numerical modelling. The art of successful modelling is to define key aspects governing prototype behaviour which can be investigated and to interpret the results to be of maximum use to engineering practice. While it is true that not all processes can be modelled, it is certainly possible to undertake meaningful investigations of buried structures and underground excavations which provide valuable data on ground movements and stress changes. The chapter will describe the difficulties associated with modelling this class of problem and discuss the achievements of some of the studies undertaken.

5.2 Buried pipes and culverts

5.2.1 Modelling considerations

All urban development involves installation of services which utilise buried pipes. Where these are large, there may be difficulties in design, particularly in ascertaining the magnitude and distribution of loads acting on the pipe. With developments in manufacturing processes, pipes can be made to be quite thin and flexible and the distribution of soil stress on such a pipe can have a major influence on its performance.

A number of different studies have been undertaken. In general, two-dimensional plane sections have been modelled and the pipes have been laid in sandy soils. The pipe may be placed in a trench with vertical or battered side slopes and backfill subsequently compacted around the pipe. In common with the prototype situation, it is very difficult to achieve uniformly com-pacted soil around a circular pipe using such a method. An alternative is to not model the trenching process but to build up the model gradually by placing the model soil in layers up to the level of the pipe invert, install the pipe and continue to place the soil. This can again lead to difficulties in achieving uniform conditions around a circular pipe with models made of dry sand, though the technique has been used successfully for tests on rectangular culverts (Hensley and Taylor, 1990). A third soil placement technique, used by Britto (1979) and Trott et al. (1984), is to fit the model container with a lid and then rotate the assembly through 90°. After removing a side wall of the container, the pipe is placed and sand poured into the box in a direction parallel to the pipe axis. This leads to a very uniform soil state around the model pipe.

It is not always possible to fabricate a model pipe from exactly the same material as the corresponding prototype. Usually, the deformation of a pipe is of key importance and consequently pipes of correctly modelled stiffness have been used; the ultimate strength of the model pipe is then often greater than for the prototype. A good example is the modelling of reinforced concrete pipes or culverts. Although micro-concrete structures can be manufactured (e.g. Hensley and Taylor, 1990), it is easier to fabricate and instrument metal models. For the example of rectangular culverts cited, a steel box section was produced which was strain gauged to monitor ring compression and longi-tudinal bending strains. The thickness of steel was chosen such that $(EI)_m = N^{-3}(EI)_p$ where N is the model scale, E is Young's modulus, I is the second moment of area per unit length for a longitudinal section and m and p refer to the model and prototype, respectively.

Pipes are usually instrumented with strain gauges. These can be placed along a circumference and in pairs such that at a particular location there is a gauge on the inside and outside faces of the pipe. Each gauge can be wired up as a quarter bridge circuit so that outer and inner circumferential strain can be determined. These can be combined to give bending moment M and hoop stress σ_h as follows:

$$M = (\varepsilon_o - \varepsilon_i) \frac{Et^2}{12(1 - v^2)} \tag{5.1}$$

$$\sigma_h = (\varepsilon_o + \varepsilon_i) \frac{E}{2(1 - v^2)} \tag{5.2}$$

where ε_o is the circumferential strain on the outer face of the pipe, ε_i is the circumferential strain on the inner face of the pipe, E is Young's modulus of the pipe material, v is Poisson's ratio of the pipe material and t is the thickness of the pipe wall.

Although this method of instrumentation is quite common, it requires considerable care if accurate results are to be obtained. Temperature changes will affect the strain readings which are to some extent compensated for when determining bending moment but can cause significant error in determining hoop stress even when nominally temperature compensated strain gauges are used. In the experiments described by Trott et al. (1984), it was found that there were significant discrepancies between a centrifuge model and the corresponding prototype when hoop stresses were compared. However, when strain measurements were compared directly, it was found that there was strong similarity between the model and prototype.

For the determination of normal stress acting on a pipe, Tohda et al. (1988, 1991a) developed a pipe section which included a number of purpose built miniature load cells. This allowed direct measurement of stress and would likely lead to a more accurate determination than one which involves manipulation of individual measurements of strain.

5.2.2 Projects

A variety of different projects have been undertaken. The Transport and Road Research Laboratory (now Transport Research Laboratory) supported research at Cambridge University into the behaviour of large-diameter, very flexible pipes. Thin-walled steel and Melinex (plastic) model pipes were tested, usually with shallow soil cover. The pipes were strain gauged as described above and the experiments were designed to investigate the extent to which pipe loading could be determined from a consideration of ring compression (Valsangkar and Britto, 1978; Britto, 1979). The results indicated that for pipes installed in wide trenches, ring compression was an adequate assumption, but for pipes installed in narrow trenches with either vertical or battered side slopes, then the line of hoop thrust fell outside the edge of the pipe. They observed that for a soil cover exceeding twice the pipe diameter there was significant reduction in load carried by the pipe due to arching.

Valsangkar and Britto (1979) report a series of tests in which a surcharge load was applied at the ground surface. They found that the effective stiffness of the pipe had a significant influence on the load reduction in the pipe due to arching. Also, they demonstrated that for flexible pipes with a shallow soil cover, pipe failure was likely to be asymmetric with buckling at the pipe shoulder. A novel analysis of centrifuge tests showed that pipe failure due to buckling could be predicted from measured deflections using the Southwell plot method (Valsangkar *et al.*, 1981).

Ko (1979) extended the research and investigated pipes with an elliptical cross-section, similar to large subway culverts. It was found that the hoop stress in the pipes was fairly uniform if the soil cover exceeded about half the vertical diameter of the section, although the line of thrust was not always within the pipe wall. Failure usually occurred by snap-through buckling at the tunnel crown unless the soil cover was very small.

Trott *et al.* (1984) describe a series of centrifuge tests and corresponding prototype tests; the experimental configurations are shown in Figure 5.1. The prototype pipe was made of steel and was 1.0 m in diameter with a wall

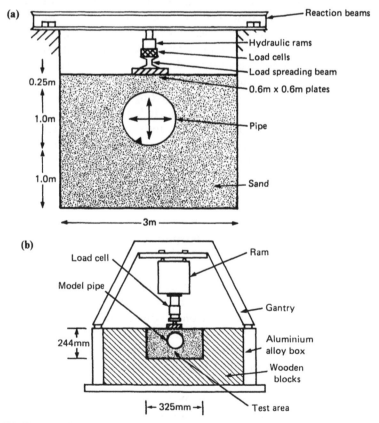

Figure 5.1 Cross-sections showing pipe installation and loading arrangements: (a) prototype pipe tested at TRRL; (b) centrifuge model. (After Trott *et al.*, 1984. Thomas Telford, with permission.)

thickness of 6 mm. It was buried in a test pit and subjected to surface strip loads placed parallel to the pipe axis and located at a number of different eccentricities. A corresponding steel model pipe approximately 109 mm in diameter was tested at 9.2 g and subjected to a similar pattern of loading as the prototype. It was found that similar strains were induced in the walls of the model and prototype pipes when subjected to corresponding loads. Figure 5.2 shows an example of the data obtained of strains in the pipe

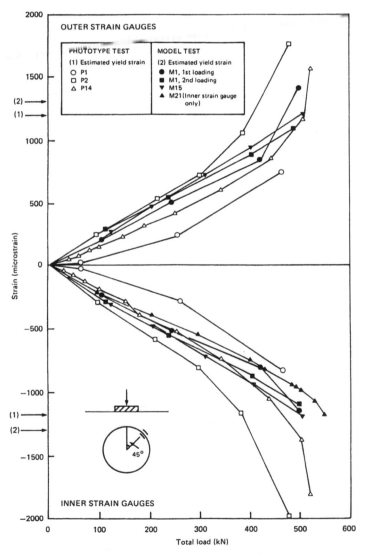

Figure 5.2 Strains induced at pipe shoulder by central strip loading for both prototype and model pipes. (After Trott *et al.*, 1984. Thomas Telford, with permission.)

shoulder due to application of a central strip load; the model loads have been converted to equivalent prototype values. The applied load–pipe deflection curves were similar although the prototype pipe appeared to have a slightly stiffer response. Also, the two pipes failed at more or less the same applied load (calculated at prototype scale) and the shapes of the distorted pipes after the test to failure were virtually identical.

Craig and Mokrani (1988) investigated the response of a model arch culvert under a central load and rolling axle loads. The experiments were designed to correspond with a full scale trial in which a heavily loaded trailer was traversed across a culvert. For the model tests, both a single-axle and a twin-axle loading unit were devised which could apply loads comparable to the prototype and which could be rolled across the model surface during centrifuge flight. There was reasonable qualitative rather than quantitative comparison between the model and prototype, mainly due to the difficulties in modelling the field construction and compaction of soil around the culvert.

Tohda et al. (1988, 1991a) report some interesting data on the behaviour of buried pipes during removal of temporary sheet piles used to support a trench during pipe placement. A number of model pipes were used which had different stiffnesses. Prior to sheet pile extraction, the pipes were fairly uniformly loaded. However, there were significant changes as the sheet piles were removed (Figure 5.3). The stress distribution around the pipe became severely non-uniform; the vertical stress on the pipes almost doubled while the horizontal stress almost halved. Significant bending strains were induced and the diametral distortion increased to about 5% of the pipe diameter for the most flexible pipe tested. The research programme also included an investigation into a number of countermeasures designed to mitigate against the adverse loading induced during sheet pile extraction.

Other research into the behaviour of buried structures has included investigations into the load transfer onto rectangular box culverts from surface loads on shallow buried culverts and from overburden loading for deep buried culverts (Hensley and Taylor, 1990; Stone et al., 1991). Model tests investigating blast induced loads have been reported by Shin et al. (1991) and Townsend et al. (1988). In these experiments instrumented micro-concrete structures were buried in sand and small explosive detonators were used to induce loads corresponding to large scale blasts.

5.3 Trenches and shafts

5.3.1 Modelling considerations

Trench and shaft excavations are necessary for many aspects of underground construction. The main difficulty in modelling trenches and shafts is in mimicking in some way the excavation process. In a centrifuge model tested

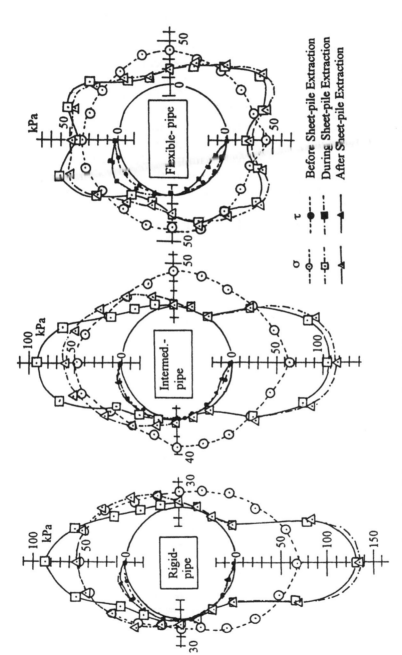

Figure 5.3 Distribution of normal and shear stresses on three model pipes of different stiffness. (After Tohda *et al.*, 1991a. Balkema, with permission.)

at many times the Earth's gravity the soil appears very heavy and it is consequently very difficult to devise small systems for excavating soil. Azevedo and Ko (1988) describe a method for modelling an excavation process by using an electric motor to raise a bag containing soil from a model and so form a sloping cut in a soil stratum. The method is simple and effective and it may be possible to develop further for modelling trench or shaft excavation processes. Craig and Yildirim (1975) devised a method of sequentially withdrawing support panels from the side of a vertical cut to model incremental excavation. Recently Kimura and his co-workers at the Tokyo Institute of Technology have devised a sophisticated system for gradual excavation in front of a retaining wall (Kimura et al., 1994). Soil is gradually scraped to one side and into a void left within the model container. Such a system could be used to investigate trench behaviour but would be difficult to adapt for modelling shaft construction.

By far the most common method used to model the excavation process has been to line a pre-formed trench or shaft with a rubber membrane which

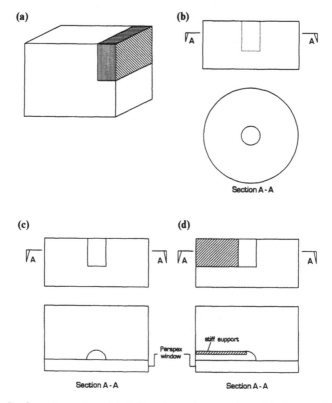

Figure 5.4 Configurations of model shaft and trench excavations: (a) plane strain trench; (b) axisymmetric shaft; (c) axisymmetric shaft with viewing on plane of symmetry; (d) simulation of trench heading.

contains a liquid. This provides a hydrostatic support to the excavation as the centrifuge reaches its operating speed. The liquid can then be drained during centrifuge flight to mimic the excavation process. This is a convenient simulation of excavation and ultimately results in realistic stress changes at the boundary of the excavation. However, the actual stress changes induced during prototype excavation are not correctly represented; for example during 'excavation', when the fluid has partly drained away, the soil below 'excavation' level is supported by a fluid pressure rather than (as in the prototype) by soil in which shear stresses can develop. This may need to be taken into account in analysis or interpretation of the results.

The common choice of liquid is zinc chloride which has an exceptionally high solubility in water. Consequently very dense fluids can be produced of comparable density to soil. With such a fluid, the vertical stress at the base of the excavation is correctly modelled and a lateral stress corresponding approximately to $K_o = 1$ is achieved. The use of zinc chloride solution was first reported by Lade et al. (1981) and was subsequently used by Kusakabe (1982), Taylor (1984), Powrie (1986) and Phillips (1986). Zinc chloride solution is highly corrosive and should be used with care (see section 4.1.3).

Configurations for trench and shaft excavations reported by Mair et al. (1984) are shown in Figure 5.4. The trench was modelled as a plane strain section which allowed observation of collapse mechanisms through the Perspex side window of a model container. The shaft can be modelled in a circular container, though it is possible to model a half section with a Perspex window coincident with the vertical plane of symmetry passing along the shaft axis. Provided the soil–window interface is well greased to limit any problems due to friction, this provides a successful means of investigating the development of deformation and collapse mechanisms.

5.3.2 Model tests

The stability and deformations due to trench excavation in clay have been studied by Craig and Yildirim (1975) and Taylor (1984). The latter work included a series of experiments investigating the long-term stability of trenches as excess pore pressures induced during excavation were allowed to dissipate. It was found that very little pore-pressure change was needed before movements accelerated and the trench collapsed; the indication was that monitoring pore-pressure changes in the field would give very little indication of imminent failure. The mechanism of collapse differed from the short-term undrained failure which was by a block of soil sliding along a 45° inclined failure plane. Instead, after a period of 'stand-up' (corresponding to a few days at prototype scale) a much narrower block of the clay fell in at the side of the excavation.

Tohda et al. (1991b) describe a series of tests on a bentonite slurry supported trench excavation in sand. Failure of the excavation was induced by raising the level of the water table in the sand. Infinitely long two-dimensional plane strain trenches were modelled as well as shorter trench panels (three-dimensional). This research had some similarities to the slurry supported excavations in clay investigated by Bolton *et al.* (1973).

Lade *et al.* (1981) conducted a series of tests on a lined circular shaft excavation in dry sand. The sand was supported by a thin Melinex tube which was strain gauged to determine the lateral thrust. The tube was lined with a rubber membrane filled with fluid. This fluid was drained during centrifuge flight to replicate the excavation process. Two fluids were used: zinc chloride

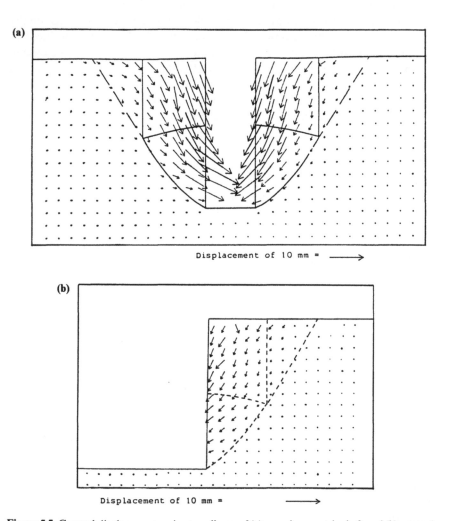

Figure 5.5 Ground displacements prior to collapse of (a) an axisymmetric shaft and (b) a trench heading. The zones of deformation predicted by plasticity solutions are indicated. (After Phillips, 1986. With permission.)

with a density corresponding to that of the dry sand was used to provide isotropic stress conditions at the shaft wall and base; and paraffin oil was used to give a much lower lateral stress corresponding to low K_0 conditions ($K_0 \simeq 1 - \sin \phi'$) though this gave poor modelling of the vertical stress at the base of the shaft. The measurements indicated that horizontal stresses acting on the shaft lining after the simulated excavation exceeded those expected from theoretical predictions and the difference was attributed in part to the flexibility of the shaft lining.

Kusakabe (1982) studied the problem of unsupported shaft excavations in clay. The results gave a valuable insight into the stability of shaft excavations and confirmed predictions based on plasticity solutions. Kusakabe (1982) also undertook a series of experiments on side-wall supported shaft excavations with the aim of establishing the depth of support required to affect the collapse mechanism and hence the stability ratio at failure. Phillips (1986) continued the work and developed models similar to the shaft but which replicated conditions at the heading of an advancing trench excavation. Figure 5.5 presents displacement vectors determined from models of an axisymmetric shaft and of a simulated trench heading. The research programme was valuable in developing new stability calculations and giving detailed insight into the pattern and zones of significant deformation due to shaft and trench excavation. Kusakabe *et al.* (1985) extended the work on shaft excavations by examining their influence during construction on adjacent shallow buried services.

5.4 Tunnels

5.4.1 Modelling considerations

A typical tunnel construction operation consists of excavation at a tunnel face with miners and machinery usually protected within a shield. The permanent tunnel lining is erected within the tailskin and as the shield is advanced, grout is usually placed between the lining and surrounding soil. The process is highly complex with potential sources of ground movement from in front of the tunnel face, around the shield and around the lining. Reproducing all details of the tunnelling process within a small scale centrifuge model would obviously be impossible and approximations need to be made which will allow key features to be investigated that are of value to engineering practice.

There are many sources of ground movement as indicated above, but if the situation when a tunnel excavation has passed a particular section is considered, the vectors of ground movement that will have developed will be more or less in the plane perpendicular to the tunnel axis. Consequently it is reasonable to assume that a plane strain model of a long tunnel section will be a good representation of tunnelling-induced movements; this is usually

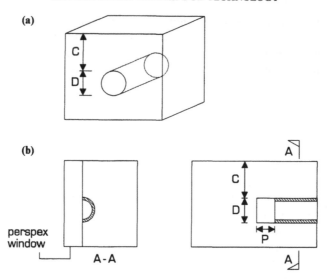

Figure 5.6 Configurations of model tunnels: (a) two-dimensional modelling of tunnel excavation; (b) longitudinal section showing modelling of tunnel heading.

referred to as a two-dimensional simulation. A typical model arrangement is shown in Figure 5.6(a). Important considerations are the tunnel diameter D which should not be too small so that difficulties in model preparation and instrumentation are avoided, the cover C which should be chosen to be within practical $C:D$ ratios of interest, and the model width. This latter point requires careful consideration to ensure that the displacement mechanism is not affected by the sidewalls of the model container. Also, the distance below the tunnel invert should not be too small, perhaps one tunnel diameter, if there is a drainage layer at the base of the model. Such a drain would act as an aquifer and could cause significant seepage-induced deformations near the tunnel which may not be representative of field conditions. In much of the modelling work undertaken, a tunnel cavity is formed in the soil which is then lined with a rubber membrane. The fluid pressure (usually compressed air) within this liner is then maintained at a value corresponding to the over-burden stress at tunnel axis level as the centrifuge speed is increased. The tunnel excavation process is simulated by reducing the tunnel support pressure, σ_T, either until collapse occurs, or to some predetermined pressure if the development of long term movements are to be studied.

Another important aspect of tunnelling is face stability during excavation which can be modelled using a longitudinal section as shown in Figure 5.6(b); this is usually referred to as a three-dimensional representation. The problem is symmetrical about a vertical plane passing along the tunnel axis and it is possible to make half section models with the Perspex sidewall of the model container located on the plane of symmetry (discussed in section 3.3). This

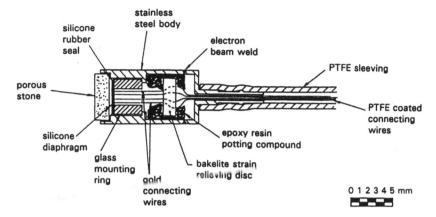

Figure 5.7 Cross-section of a Druck PDCR81 miniature pore-pressure transducer.

allows movements above and towards the model tunnel heading to be observed. A stiff support can be used to represent the lining and there may be a length of structurally unsupported soil left near the tunnel face. There are similar considerations on geometry as before and the structurally unsupported length of heading, P, can be varied to be within the range encountered in practice. Again a rubber lining membrane is used to contain the tunnel air support pressure.

A major programme of research on tunnels was supported by the then Transport and Road Research Laboratory (O'Reilly *et al.*, 1984) and undertaken at Cambridge University during the 1970s and early 1980s. The later work included effective stress modelling with detailed measurement of pore-pressure changes induced during tunnel excavation. Extensive use was made of the miniature pore pressure transducers, type PDCR81, manufactured by Druck Ltd. A cross-section of the transducer is shown in Figure 5.7. These transducers have proved to be invaluable and have been used in many centrifuge test programmes.

5.4.2 Tunnels in sand

A series of plane strain (two-dimensional) models of tunnels in dry sand are described by Potts (1976) and Atkinson *et al.* (1977). The general outline of the model corresponded to that shown in Figure 5.6(a). A model tunnel diameter of 60 mm was tested with $C:D$ ratios in the range 0.4–2.4. A centrifuge acceleration of 75 g was used so the model tunnel corresponded to a prototype tunnel of 4.5 m diameter, i.e. comparable to that used for many underground transit systems. The model container was 360 mm wide which was probably just sufficient to avoid boundary side-wall effects even for the deepest tunnels tested.

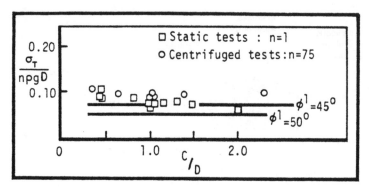

Figure 5.8 Tunnel pressures at collapse for unlined tunnels in dense sand. (After Atkinson *et al.*, 1977. Morgan-Grampian, with permission.)

The centrifuge tests followed on from a series of $1\,g$ laboratory tests on tunnels of a similar geometry. It was observed that the minimum tunnel support pressure required to prevent collapse was very small and, except for cases where $C:D$ was very small, was independent of the magnitude of soil cover (Figure 5.8). Following observations of ground deformation at failure, a new collapse mechanism was devised which explained the experimental results (Atkinson *et al.*, 1975; Atkinson and Potts, 1977) and the predictions are shown by the thick horizontal lines in Figure 5.8. It is interesting to note that for back-analysis of the centrifuge test data, a lower angle of friction needs to be used than for the static $1\,g$ experiments. This is to be expected due to the greater and more realistic effective stresses experienced in the centrifuge models. A higher stress level has the effect of reducing the component of dilation during shear and consequently the peak strength is lower.

Investigations on tunnel face stability are described by Chambon (1991). The model geometry was similar to that indicated in Figure 5.6(b). The conclusions reached were very similar to those for the plane section models described above, i.e. that only a small tunnel support pressure was needed to prevent collapse of the tunnel face. The minimum support pressure was independent of soil cover provided $C:D > 1$. The propagation and extent of the failure mechanism was clarified by including thin layers of coloured sand within the soil model. There was no back-analysis of the failure but it is likely that a similar mechanism to that devised by Atkinson and Potts (1977) for the plane strain models would be appropriate.

Güttler and Stöffers (1988) describe an interesting series of experiments in which support for the tunnel cavity was provided by a brittle lining. The cast lining was made from a mixture of gypsum and kaolin. The tests were designed to be representative of shotcrete lined tunnels (often referred to as the New Austrian Tunnelling Method). An interesting technique used in these experiments was crack detection in the brittle lining using strips of conductive paint. The reduction in overburden stress transferred to the lining was shown to be a function of the lining stiffness. In these experiments, the load reduction was limited due to the relatively high stiffness of the lining and the initial 'good fit' between the circular lining and the uniformly prepared soil; collapse generally followed the formation of a shear failure in the lining at about springing level.

König et al. (1991) investigated the stress redistribution that may occur during tunnel and shaft construction in sand. The model consisted of partly supported tunnels or shafts in which the structural lining did not extend to the limit of excavation. The structurally unsupported zone was stabilised by air pressure acting within a flexible lining membrane. It was found that in the dry sand only a small support pressure was needed to maintain stability. However, there was a significant load transfer to the lining closest to the excavation face due to arching and stress redistribution in the soil. This stress concentration was found to be less significant for the more flexible linings tested.

5.4.3   Tunnels in silt

The tunnel tests in sand used dry soil and so avoided any problems due to transient or steady seepage flow to the tunnel. In order to investigate these problems, a series of model tests in saturated silt were undertaken by Taylor (1979) and are described briefly by Schofield (1980). A silica rock flour with a grading in the silt range was used since it had both a low compressibility such that any transient seepage (consolidation) effects would be of very short duration, and a moderately low permeability so that any seepage induced problems would not develop too rapidly to observe. A two-dimensional plane strain model configuration was used, i.e. as Figure 5.6(a). The test proceeded by reducing the tunnel air support pressure to a value which would permit seepage flow to the tunnel but which would avoid immediate failure. In these tests the water table was maintained at the ground surface and so seepage could occur provided $\sigma_T < \gamma_w(C+D)$ (where γ_w is the bulk unit weight of water). The pressure head driving the seepage flow is then the difference between the hydrostatic pore pressure at the tunnel invert and the tunnel air pressure.

In the experiments using silt, it was found that the pore pressures quickly reached equilibrium values consistent with steady seepage flow towards the tunnel. There was a time period during which little change in pore pressure or surface settlement occurred and this was later referred to as 'stand-up' time. After this period, there was a significant increase in settlement and as the tunnel failed, so erratic transient pore pressures were monitored reflecting the shearing in the soil. Two photographs taken during the test are shown in Figure 5.9. The high-speed flash of the photographic system cast a shadow of the tunnel plug in the front Perspex face of the model container. Figure 5.9(a)

(a)

Figure 5.9 Photographs from a centrifuge test of an unlined tunnel in silt: (a) during the 'stand-up' period; (b) just after the onset of failure. (After Taylor, 1979.)

was taken at a stage towards the end of the 'stand-up' period and silvered plastic marker beads pressed into the face of the model show evidence of some settlement above the tunnel. Figure 5.9(b) is a photograph taken just after the onset of failure. The marker beads indicate movement towards the tunnel and the high shear strains caused the silt to liquefy in the region to the side of the tunnel and the marker beads became obscured in that zone. Back-analysis of the failure indicated that the same period of 'stand-up' was consistent with that needed for the pool of water collecting in the tunnel to reach a depth at which sidewall erosion of the tunnel could occur leading to imminent instability.

(b)

5.4.4   Tunnels in clay

Considerable advances in the understanding of tunnel behaviour have followed on from centrifuge studies and a summary is included in Mair *et al.* (1984). Mair (1979) undertook a number of series of centrifuge tests aimed at clarifying the factors affecting tunnel stability (Kimura and Mair, 1981) and investigating the ground deformations due to tunnel construction (Mair *et al.*, 1981). The different objectives of the various test series lead to slightly different modelling techniques being used. For investigation of stability, analysis is often simplified if the soil has a constant shear strength (Davis *et al.*, 1980). Consequently, for these studies the clay was consolidated in the laboratory to a constant vertical effective stress after which the tunnel was cut in the soil sample prior to undertaking the centrifuge test. Stability of the model tunnel during centrifuge speed increase was assured by increasing the air pressure in the rubber bag lining the tunnel to correspond to the overburden stress at tunnel axis level. After reaching the test acceleration, the tunnel support pressure was reduced to zero in a period of about 2–3 min. Thus there was insufficient time for the sample to reconsolidate during centrifuge flight and so the effective stress and hence shear strength of the clay should remain fairly constant with depth. Another test series was concerned not only with stability but also with the spread of deformations around a tunnel as the support pressure was reduced. For these experiments, the stress–strain behaviour of the ground was important and the clay was reconsolidated to effective stress equilibrium in centrifuge flight. The machine was then stopped, the tunnel cut and the centrifuge restarted, again using an air support pressure in the tunnel to prevent collapse as the operating acceleration was reached. It is important when using such a procedure to keep the stop–restart cycle to a minimum period to avoid effective stress changes within the clay. All the models were made from kaolin clay with a shear strength in the region 25–35 kPa. The model tunnel was usually 60 mm in diameter tested at 75 g.

The first tests were conducted using the plane strain configuration (Figure 5.6a) and were a parametric study designed to investigate the influence of tunnel geometry and soil conditions on tunnel stability. By using dimensional analysis, it can be shown that for the plane strain model, the tunnel support pressure at collapse, σ_{Tc} depends on the dimensionless groups:

$$\frac{\sigma_{\mathrm{Tc}}}{c_{\mathrm{u}}} = f_1 \left[\frac{N\gamma D}{c_{\mathrm{u}}}, \frac{C}{D} \right] \tag{5.3}$$

where N is the centrifuge acceleration factor and γ is the bulk unit weight of the clay (at 1 g) having an undrained shear strength c_{u}. The observed values of

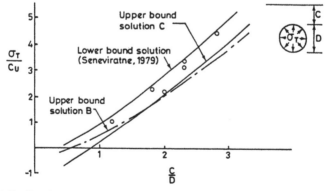

Figure 5.10 Predicted and observed tunnel support pressures at collapse of plane section model tunnels. (After Mair, 1979.)

σ_{Tc} for a number of centrifuge tests in which the cover-to-diameter ($C:D$) ratio was varied are shown in Figure 5.10; for these tests, $N\gamma D/c_u$ was 2.6. Also shown on the figure are results from upper and lower bound plasticity solutions (Davis *et al.*, 1980). Limit analyses of this type are of immense value and the solutions presented were developed in the light of the failure mechanisms observed from the centrifuge tests. As can be seen, the upper and lower bound solutions are very similar and the centrifuge tests have played an important role in validating the analyses.

A series of tests was also undertaken in which the stability of a tunnel heading was investigated. For this situation, as shown in Figure 5.6(b), dimensional analysis indicates that;

$$\frac{\sigma_{\text{Tc}}}{c_u} = f_2\left[\frac{N\gamma D}{c_u}, \frac{C}{D}, \frac{P}{D}\right] \tag{5.4}$$

Results from the centrifuge tests and theoretical analysis showed that the expression can be simplified as:

$$T_c = f_3\left[\frac{C}{D}, \frac{P}{D}\right] \tag{5.5}$$

where the stability ratio at collapse T_c is defined as:

$$T_c = \frac{N\gamma\left(C + \dfrac{D}{2}\right) - \sigma_{\text{Tc}}}{c_u} \tag{5.6}$$

There is some analogy between T_c used by tunnel engineers and the foundation bearing capacity factor N_c (Atkinson and Mair, 1981). The centrifuge test results led to the chart shown in Figure 5.11 which gives a clear insight

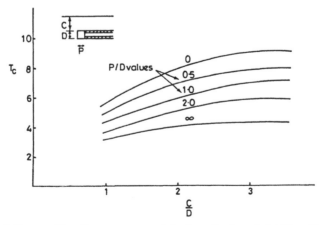

Figure 5.11 Influence of heading geometry and depth on the tunnel stability ratio at collapse. (After Mair, 1979.)

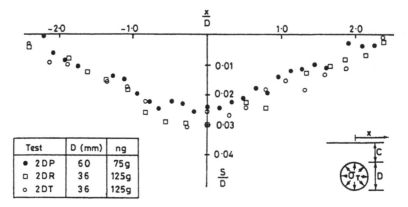

Figure 5.12 Surface settlement profiles above geometrically similar model tunnels tested at different scales; $C:D = 1.67$, $\sigma_T = 92\,\text{kPa}$. (After Mair, 1979.)

into how the geometry of a tunnel heading influences stability. T_c tends to a fairly constant value at large $C:D$ (> 3) and there is clearly significant benefit in providing support close to the tunnel face (i.e. minimising $P:D$). The results represent a major advance in the understanding of the stability of tunnels during construction.

In tunnel design it is important to be able to predict ground deformations due to construction. Tests were undertaken using plane strain (two-dimensional) models in which the clay was first brought into effective stress equilibrium so that realistic patterns of deformation would be observed. Also, within the test programme, geometrically similar tunnels were tested at different accelerations such that they corresponded to the same prototype. This modelling of models exercise is useful to validate the centrifuge scaling laws and so give credibility to the observed movements. An example of the settlement troughs observed above geometrically similar tunnels tested at different accelerations is shown on Figure 5.12. The results have been normalised to demonstrate their close agreement. An important aspect of deformations due to tunnelling is the width of the settlement trough. In general, the centrifuge tests demonstrated that for a given tunnel depth, the trough width was independent of the degree of support. The centrifuge tests were significant in that they implied that the spread of surface settlements due to tunnelling is independent of the method of construction.

In assessing tunnel behaviour prior to collapse, the use of load factor is beneficial. Load factor, LF, is the reciprocal of factor of safety, F, and can be defined as:

$$LF = \frac{N\gamma\left(C + \dfrac{D}{2}\right) - \sigma_T}{N\gamma\left(C + \dfrac{D}{2}\right) - \sigma_{Tc}} \tag{5.7}$$

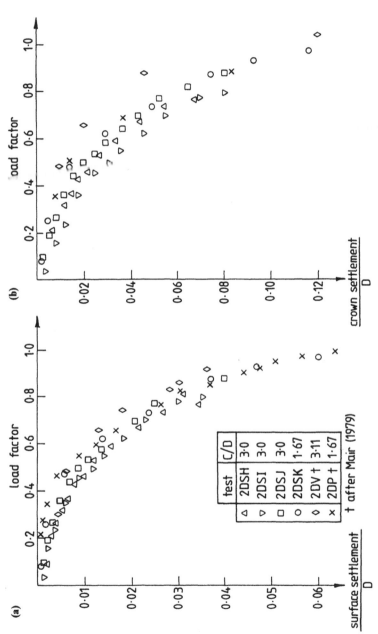

Figure 5.13 Observed variations of maximum surface settlement (a) and tunnel crown settlement (b) with load factor. (After Taylor, 1984.)

where σ_T is the tunnel support pressure and is greater than σ_{Tc} to avoid collapse. Thus LF = 0 when σ_T corresponds to the overburden stress at tunnel axis level and increases to LF = 1 at failure with $\sigma_T = \sigma_{Tc}$. Values of maximum surface settlement, S_s, and tunnel crown settlement, S_c, are shown plotted against LF in Figure 5.13 for a number of centrifuge tests on tunnels

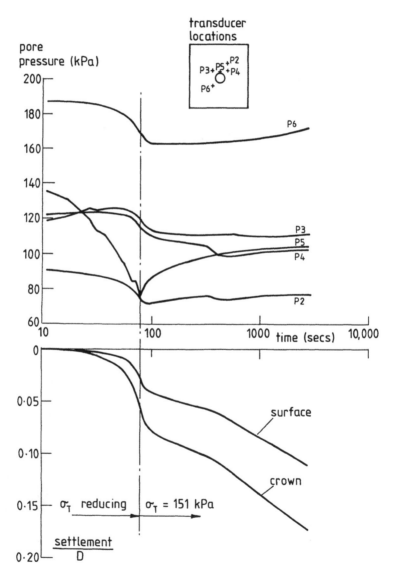

Figure 5.14 Variation of pore water pressures and settlements with time observed in a model tunnel test ($C{:}D = 3.0$, load factor = 0.76). (After Taylor, 1984.)

at different depths. The results indicate that settlement and load factor are interrelated which is useful for tunnel design (Mair *et al.*, 1984).

The deformations considered above relate to the undrained event of removing support as a tunnel is excavated. This is often the major component of movement but sometimes time-dependent deformations (often referred to as 'squeeze') can be relevant. Centrifuge tests investigating this type of event were undertaken by Taylor (1984) and a typical set of results is presented in Figure 5.14.

In the first phase of the experiment, the tunnel pressure is reduced quickly causing a reduction in pore pressure and some ground deformations. In the subsequent phase, with the tunnel pressure maintained at a constant value, it was observed that settlements continued as the pore pressures gradually increased. The almost linear increase of settlement on a logarithmic time scale could lead to the erroneous conclusions that continuing ground movements were associated with some viscous behaviour of the clay. However, the rate at which these deformations occur and their association with pore pressure changes indicates that time-dependent movements near tunnels are uniquely related to effective stress changes in clay and this has been confirmed using finite element analysis (Taylor, 1984; De Moor and Taylor, 1991).

5.5 Summary

From the above discussion it can be seen that modelling excavations and excavation processes can be difficult, but that it is possible to make sensible approximations so that valuable and meaningful data can be obtained. It is possible to obtain reasonable assessments of collapse and strong links have been made with analyses based on plasticity solutions. Patterns of deformation have been observed and it has been found that, in general, these are consistent with the collapse mechanisms. The class of research described has been of immense value in developing empirical design procedures, and also in identifying risk zones of significant and damaging movements.

References

Atkinson, J.H. and Potts, D.M. (1977) The stability of a shallow circular tunnel in cohesionless soil. *Géotechnique*, **27** (2), 203–215.

Atkinson, J.H. and Mair, R.J. (1981) Soil mechanics aspects of soft ground tunnelling. *Ground Eng.* **14** (5), 20–28.

Atkinson, J.H., Brown, E.T. and Potts, D.M. (1975) Collapse of shallow unlined circular tunnels in dense sand. *Tunnels and Tunnelling*, **7** (3), 81–87.

Atkinson, J.H., Potts, D.M. and Schofield, A.N. (1977) Centrifugal model tests on shallow tunnels in sand. *Tunnels and Tunnelling*, **9** (1), 59–64.

Azevedo, R.F. and Ko, H.-Y. (1988) In-flight centrifuge excavation tests in sand. *Centrifuge '88, Paris*, pp. 119–124. Balkema, Rotterdam.

Bolton, M.D., English, R.J., Hird, C.C. and Schofield, A.N. (1973) Ground displacements in centrifugal models. *Proc. 8th Int. Conf. Soil Mech. Found. Eng., Moscow*, Vol. 1, pp. 65–70.

Britto, A.M. (1979) Thin Walled Buried Pipes. PhD Thesis, Cambridge University.

Chambon, P., Corté, J.-F., Garnier, J. and König, D. (1991) Face stability of shallow tunnels in granular soils. *Centrifuge '91, Boulder, Colorado*, pp. 99–105. Balkema, Rotterdam.

Craig, W.H. and Mokrani, A. (1988) Effect of static line and rolling axle loads on flexible culverts buried in granular soil. *Centrifuge '88, Paris*, pp. 385–394. Balkema, Rotterdam.

Craig, W.H. and Yildirim, S. (1975) Modelling excavations and excavation processes. *Proc. 6th Eur. Conf. Soil Mech. Found. Eng., Vienna*, Vol. 1, pp. 33–36.

Davis, E.H., Gunn, M.J., Mair, R.J. and Seneviratne, H.N. (1980). The stability of shallow tunnels and underground openings in cohesive material. *Géotechnique*, **30** (4), 397–416.

De Moor, E.K. and Taylor, R.N. (1991) Time dependent behaviour of a tunnel heading in clay. *Proc. 7th Int. Conf. Computer Methods and Advances in Geomechanics, Cairns, Queensland*, Vol. 2, pp. 1455–1460. Balkema, Rotterdam.

Güttler, U. and Stöffers, U. (1988) Investigation of the deformation collapse behaviour of circular lined tunnels in centrifuge model tests. In *Centrifuges in Soil Mechanics* (eds W.H. Craig, R.G. James and A.N. Schofield), pp. 183–186. Balkema, Rotterdam.

Hensley, P.J. and Taylor, R.N. (1990) Centrifuge Modelling of Rectangular Box Culverts. Geotechnical Engineering Research Centre Report GE/90/13, City University, London.

Kimura, T. and Mair, R.J. (1981) Centrifugal testing of model tunnels in soft clay. *Proc. 10th Int. Conf. Soil Mech. Found. Eng., Stockholm, June 1981*, Vol. 1, pp. 319–322. Balkema, Rotterdam.

Kimura, T., Takemura, J., Hiro-oka, A., Okamura, M. and Park, J. (1994) Excavation in soft clay using an in-flight excavator. *Centrifuge '94* (eds C.F. Leung, F.H. Lee and T.S. Tan), pp. 649–654. Balkema, Rotterdam.

Ko, H.-Y. (1979) Centrifuge Model Tests of Flexible, Elliptical Pipes Buried in Sand. Cambridge University Engineering Department Technical Report CUED/D—Soils TR 67.

König, D., Güttler, U. and Jessberger, H.L. (1991) Stress redistributions during tunnel and shaft constructions. *Centrifuge '91, Boulder, Colorado*, pp. 129–135. Balkema, Rotterdam.

Kusakabe, O. (1982) Stability of Excavations in Soft Clay. PhD Thesis. Cambridge University.

Kusakabe, O., Kimura, T., Ohta, A. and Takagi, N. (1985) Centrifuge model tests on the influence of axisymmetric excavation on buried pipes. *Proc. 3rd Int. Conf. Ground Movements and Structures, Cardiff*, pp. 113–128. Pentech Press, London.

Lade, P.V., Jessberger, H.L., Makowski, E. and Jordan, P. (1981) Modelling of deep shafts in centrifuge tests. *Proc. 10th Int. Conf. Soil Mech. Found. Eng., Stockholm*, Vol. 1, pp. 683–691. Balkema, Rotterdam.

Mair, R.J. (1979) Centrifugal Modelling of Tunnel Construction in Soft Clay. PhD Thesis, Cambridge University.

Mair, R.J., Gunn, M.J. and O'Reilly, M.P. (1981) Ground movements around shallow tunnels in soft clay. *Proc. 10th Int. Conf. Soil Mech. Found. Eng., Stockholm*, Vol. 1, pp. 323–328. Balkema, Rotterdam.

Mair, R.J., Phillips, R., Schofield, A.N. and Taylor, R.N. (1984) Applications of centrifuge modelling to the design of tunnels and excavations in clay. *Symp. Application of Centrifuge Modelling to Geotechnical Design*, pp. 357–380. Balkema, Rotterdam.

O'Reilly, M.P., Murray, R.T. and Symons, I.F. (1984) Centrifuge modelling in the TRRL research programme on ground engineering. *Symp. Application of Centrifuge Modelling to Geotechnical Design*, pp. 423–440. Balkema, Rotterdam.

Phillips, R. (1986) Ground Deformation in the Vicinity of a Trench Heading. PhD Thesis, Cambridge University.

Potts, D.M. (1976) Behaviour of Lined and Unlined Tunnels in Sand. PhD Thesis, Cambridge University.

Powrie, W. (1986) The Behaviour of Diaphragm Walls in Clay. PhD Thesis, Cambridge University.

Schofield, A.N. (1980) Cambridge geotechnical centrifuge operations. *Géotechnique*, **30** (3), 227–268.

Seneviratne, H.N. (1979) Deformations and Pore Pressure Dissipation around Shallow Tunnels in Soft Clay. PhD Thesis, Cambridge University.

Shin, C.J., Whittaker, J.P., Ko, H.-Y. and Sture, S. (1991) Modelling of dynamic soil-structure interaction phenomena in buried conduits. *Centrifuge '91, Boulder, Colorado*, pp. 457–463. Balkema, Rotterdam.

Stone, K.J.L., Hensley, P.J. and Taylor, R.N. (1991) A centrifuge study of rectangular box culverts. *Centrifuge '91, Boulder, Colorado*, pp. 107–112. Balkema, Rotterdam.

Taylor, R.N. (1979) 'Stand-up' of a Model Tunnel in Silt. MPhil Thesis, Cambridge University.

Taylor, R.N. (1984) Ground Movements Associated with Tunnels and Trenches. PhD Thesis, Cambridge University.

Tohda, J., Mikasa, M. and Hachiya, M. (1988) Earth pressure on underground rigid pipes: centrifuge model tests and FEM analysis. *Centrifuge '88, Paris*, pp. 395–402. Balkema, Rotterdam.

Tohda, J., Yoshimura, H., Ohi, K. and Seki, H. (1991a) Centrifuge model tests on several problems of buried pipes. *Centrifuge '91, Boulder, Colorado*, pp. 83–90. Balkema, Rotterdam.

Tohda, J., Nagura, K., Kawasaki, K., Higuchi, Y., Yagura, T. and Yano, H. (1991b) Stability of slurry trench in sandy ground in centrifuged models. *Proc. Centrifuge '91, Boulder, Colorado*, pp. 75–82. Balkema, Rotterdam.

Townsend, F.C., Tabatabai, H., McVay, M.C., Bloomquist, D. and Gill, J.J. (1988) Centrifugal modelling of buried structures subjected to blast loadings. *Proc. Centrifuge '88, Paris*, pp. 473–479. Balkema, Rotterdam.

Trott, J.J., Symons, I.F. and Taylor, R.N. (1984) Loading tests to compare the behaviour of full scale and model buried steel pipes. *Ground Eng.*, 17 (6), 17–28.

Valsangkar, A.J. and Britto, A.M. (1978) The Validity of Ring Compression Theory in the Design of Flexible Buried Pipes. Transport and Road Research Laboratory Supplementary Report SR 440.

Valsangkar, A.J. and Britto, A.M. (1979) Centrifuge Tests of Flexible Circular Pipes Subjected to Surface Loading. Transport and Road Research Laboratory Supplementary Report SR 530.

Valsangkar, A.J., Britto, A.M. and Gunn, M.J. (1981) Application of the Southwell plot method to the inspection and testing of buried flexible pipes. *Proc. Inst. Civil Engineers*, Part 2, 71, March, pp. 63–82.

6 Foundations

O. KUSAKABE

6.1 Introduction

Modelling of foundation behaviour is the main focus of many centrifuge studies. A wide range of foundations have been used in practical situations, including spread foundations, pile foundations and caissons. The main objectives of centrifuge modelling for foundation behaviour are to investigate: (i) load–settlement curves from which yield and ultimate bearing capacity as well as stiffness of the foundation may be determined; (ii) the stress distribution around and in foundations, by which the apportionment of the resistance of the foundation to bearing load and the integrity of the foundation may be examined; and (iii) the performance of foundation systems under working loads as well as extreme loading conditions such as earthquakes and storms.

The construction sequence has relatively little influence on the behaviour of shallow foundations, while there exists a significant effect of installation method on deep foundation behaviour. For modelling purposes, therefore, it is appropriate to distinguish clearly between shallow and deep foundations. In this respect, centrifuge modelling of deep foundations needs more careful consideration of the stress changes in the model soil during installation. Development of systems simulating the installation process also forms an important part of modelling of deep foundations.

6.2 Shallow foundations

Shallow foundations may be classified in terms of the shape of footing, loading condition and ground conditions as shown in Table 6.1. The foundation problems which have been examined by centrifuge tests are identified as well as the problems which still need experimental verification of theoretical analyses. Also, it can be seen from the table the extent to which centrifuge modelling has been utilized extensively in various kinds of foundation problems, which demonstrates the usefulness of centrifuge modelling in the area of foundation engineering.

Table 6.1 Bearing capacity problems for shallow footings

Shape of footing	Strip			Circular			Rectangular		
Loading condition*	V	E	I	V	E	I	V	E	I
Single									
uniform layer, c, φ	T/C	T/C	T/C	T/C	T		T/C	T	
NC	T/C	T		T			T		
anisotropic sand	T/C			C					
anisotropic clay	T/C								
NC with anisotropy	T/C								
Multi-layer									
two layers, uniform, c, φ	T								
two layers, uniform clay	T			T			T		
two layers, NC with anisotropy	T								
sand overlaying clay	T/C			T/C			T		
clay sandwiched by stiff layers	T/C			T/C			T		
three layers	T								
Slope									
uniform c, φ	T/C	T							

*V = vertical, E = eccentric, I = inclined, T = theoretical solution(s), C = centrifuge studies.

6.2.1 Brief review of bearing capacity formulae

The well-known Terzaghi bearing capacity formula for a shallow strip footing on uniform soil is expressed as:

$$\frac{Q}{B} = q = cN_c + \gamma D N_q + \gamma \frac{B}{2} N_\gamma \qquad (6.1)$$

where Q is the applied load per unit length, B is the width of footing, D is the depth of embedment, γ is the unit weight of soil and N_c, N_q, N_γ are the bearing capacity factors which are a function of a constant angle of friction ϕ'.

For a footing on or in purely 'cohesive' soils, i.e. under undrained loading conditions, equation (6.1) reduces to:

$$q = c_u N_c + \gamma D \qquad (6.2)$$

where c_u is the undrained shear strength. Equation (6.2) is further simplified for a surface footing as:

$$q = c_u N_c \qquad (6.3)$$

Equation (6.3) indicates that the bearing capacity of a shallow footing on a clay having a constant strength with depth, is only dependent on the undrained strength and is independent of the footing size. In many practical situations, however, the strength of clay layer increases linearly with depth and is often expressed as:

$$c(z) = c_0 + kz \tag{6.4}$$

where c_0 = shear strength at the surface of the soil layer, k = rate of increase of shear strength with depth, and z = depth.

The bearing capacity of a footing on normally consolidated clay (NC clay) is then expressed by:

$$q = c_0 N_{co} \tag{6.5}$$

where N_{co} = the bearing capacity factor which is then a function of kB/c_0. In cases where the shear strength at the surface is zero ($c_0 = 0$), equation (6.5) is written as:

$$q = \frac{kB}{4} \tag{6.6}$$

If the thickness of the clay layer beneath a footing is thin relative to its width, a failure mode by squeezing becomes critical and the bearing capacity is then written in the form:

$$q = c_u \left(N_c + \alpha \frac{B}{H} \right) \tag{6.7}$$

in which H is the thickness of clay layer, and α is a constant.

In contrast, for cohesionless soils, i.e. for drained loading conditions on soils with no true cohesion, equation (6.1) reduces to:

$$q = \gamma D N_q + \gamma \frac{B}{2} N_\gamma \tag{6.8}$$

Equation (6.8) may be rewritten in a slightly different form of:

$$q = \gamma \frac{B}{2} N_{\gamma q} \tag{6.9}$$

where $N_{\gamma q}$ = the combined bearing capacity factor. Equation (6.8) is further reduced to:

$$q = \gamma \frac{B}{2} N_\gamma \tag{6.10}$$

for a surface footing. It can be seen from equation (6.10) that the bearing capacity of footings on cohesionless soil is a function of footing size and increases linearly with an increase in footing width B.

For eccentric loading conditions, Meyerhof's hypothesis of effective width is usually adopted as:

$$B' = B - 2e \tag{6.11}$$

where B' = effective width, and e = eccentricity of the load.

Under general combined loading conditions, yield surfaces are constructed on axes of vertical load–horizontal load (V–H space) and of vertical load and moment load (V–M space).

Real footings are, of course, not always of a strip shape. The effect of footing shape is taken into account by introducing shape factors. Equation (6.1) may be rewritten as:

$$q = s_c c N_c + s_q \gamma D N_q + s_\gamma \gamma \frac{B}{2} N_\gamma \tag{6.12}$$

6.2.2 Shallow foundations on cohesionless soils

6.2.2.1 Modelling of models. Attempts at modelling of models have been made for cases of vertical bearing capacity of footings on dry sand (equations 6.8 and 6.9). Ovesen (1980) carried out a test programme of concentric loading tests of circular footings on a uniform dry diluvial sand ($e = 0.565$, $D_{50} = 0.3$–0.6 mm), varying the combination of the diameter of footing and acceleration as illustrated on Figure 6.1.

The tests results are presented in the form of dimensionless load–settlement curves in Figure 6.2. It should be noted that all the tests correspond to a prototype diameter of $d_p = 1.0$ m. The centrifuge scaling laws are validated by these results. Similar studies have been undertaken by Mikasa *et al.* (1973), Ovesen (1975), Yamaguchi *et al.* (1977), Corté *et al.* (1988), Pu and Ko (1988) and Kutter *et al.* (1988b).

Figure 6.3 summarizes the results of modelling of models by plotting bearing capacity against acceleration for various sands, from which it can be concluded that modelling of models holds. Another set of data from the same

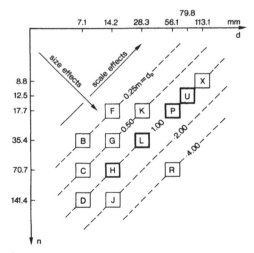

Figure 6.1 Test programme for modelling of models. (After Ovesen, 1980. British Geotechnical Society, with permission.)

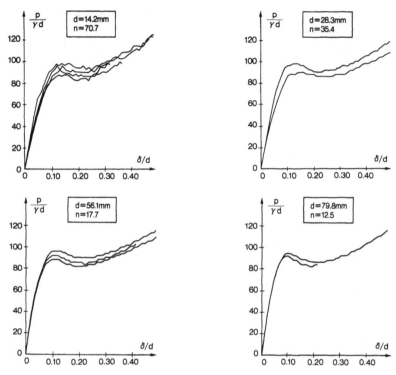

Figure 6.2 Normalized load-settlement curves for modelling of models. (After Ovesen, 1980. British Geotechnical Society, with permission.)

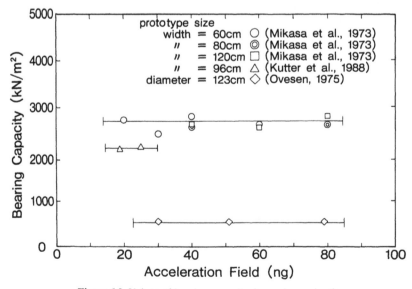

Figure 6.3 Values of bearing capacity for various g levels.

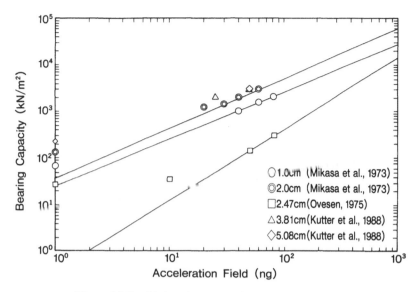

Figure 6.4 Double log plots of bearing capacity vs. g level.

authors is plotted in terms of bearing capacity and acceleration using double logarithmic scales in Figure 6.4, from which it can be seen that the angle of internal friction in the bearing capacity formula should not be a constant, as was formulated in equation (6.1).

It should be noted that the first conclusion is limited to approximately a threefold range of scale factor. The essence of the second conclusion was already demonstrated experimentally in the 1940s and will be discussed later.

The most extreme case of modelling of models is to compare full-scale field test results directly with centrifuge tests using a corresponding model. Three European centrifuge laboratories have cooperated on this subject (Corté et al., 1988; Bagge et al., 1989). The full-scale tests were carried out in a test pit and load tests were performed using a flat circular footing of 1.6 m diameter on saturated compacted Fontainebleau sand. Figure 6.5 shows the load deflection curve found in the full scale test and curves from model tests determined in the three laboratories. The model ground was made by means of air pluviation and the diameter of model footing was 56.6 mm. It is noticed that the loads agree reasonably at maximum deflection. However, the initial modulus of sub-grade reaction observed at full scale is larger than any of the model test results by a factor of 2–3.

A similar project was conducted on an undisturbed granular soil (Fujii et al., 1988). They compared large-scale in situ field loading tests of a surface footing on a slope made in a pumice flow deposit with corresponding centrifuge tests using undisturbed samples obtained from the loading test site.

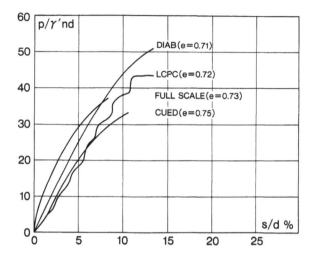

Figure 6.5 Comparison of load deflection curves between full-scale test and centrifuge tests. (After Bagge *et al.*, 1989.)

Figure 6.6 is the comparison of the load–settlement curves, from which they concluded that the centrifuge tests could predict ultimate bearing capacity fairly accurately, but overestimated settlements by a factor of 2–3.

6.2.2.2 Scale effect. It has been observed that the bearing capacity factor N_γ decreases with an increase in footing width B. This phenomenon implies that bearing capacity does not increase linearly with the footing width, and therefore that equations (6.8) and (6.9) are not exact. The phenomenon was widely recognized among geotechnical engineers and was discussed at length by de Beer (1963). This phenomenon may be, thus, termed de Beer's scale effect of the footing. Since then, the subject has attracted a number of research workers, in particular centrifuge modellers (Yamaguchi *et al.*, 1977; Terashi *et al.*, 1984; Kutter *et al.*, 1988b; Kusakabe *et al.*, 1991).

It can be seen from the data presented by Yamaguchi *et al.* (1977) shown in Figure 6.7 that N_γ generally decreases with an increase in the parameter $\gamma B/E$. The decrease in N_γ with $\gamma B/E$ given by both de Beer (1967) and Vesic (1963) are sharper than in those data. Kimura *et al.* (1985) added further experimental data and demonstrated that the scale effect becomes less marked with a decrease in the relative density of the soil.

Figure 6.8 summarizes other data available on the scale effect of N_γ on dry sand. The trend may be conveniently expressed in a form of a power function as:

$$N_\gamma = N_{\gamma o}\left(\frac{B}{B_o}\right)^{-\alpha} \tag{6.13}$$

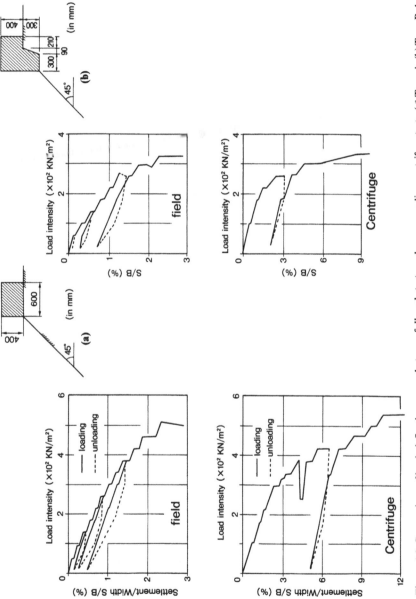

Figure 6.6 Comparison of load–deflection curves between full-scale tests and corresponding centrifuge tests. (a) Type A, (b) Type B-4. (After Fujii *et al.*, 1988. Balkema, with permission.)

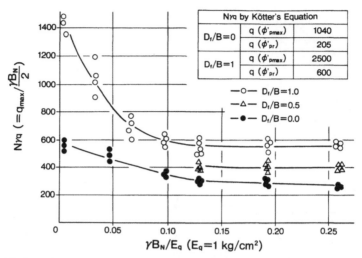

Figure 6.7 Centrifuge data for de Beer's scale effect. (After Yamaguchi *et al.*, 1977. Japanese Society of Soil Mechanics and Foundation Engineering, with permission.)

Kusakabe *et al.* (1991) examined the scale effect in terms of footing shape by using aspect ratios ranging from 1 to 7. The data indicated that the degree of the scale effect seems to be dependent on the geometry of the footing: the bearing capacity of circular footings decreases more sharply with increasing

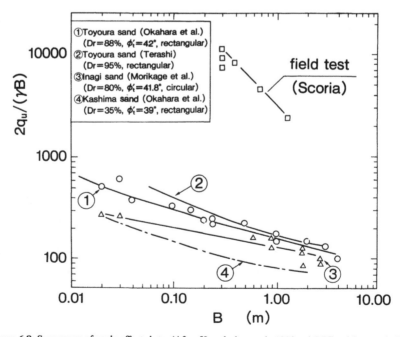

Figure 6.8 Summary of scale effect data. (After Kusakabe *et al.*, 1992a. ASCE, with permission.)

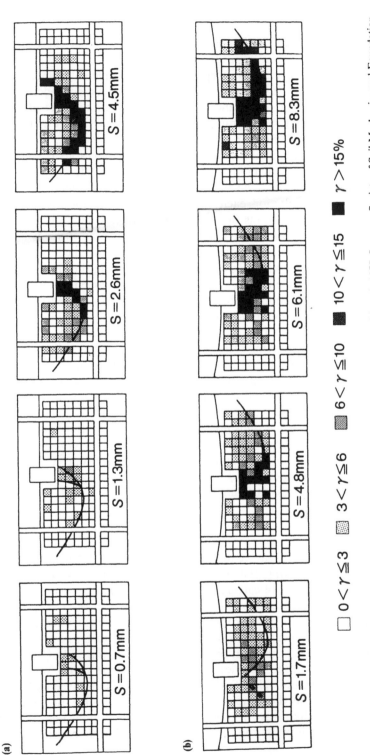

Figure 6.9 Strain development with settlement in loading tests. (a) 1 g, (b) 30 g. (After Yamaguchi *et al.*, 1977. Japanese Society of Soil Mechanics and Foundation Engineering, with permission.)

the width of the footing. This implies that there is also a scale effect of shape factor as well.

The scale effect of N_γ may be interpreted from two different points of view: progressive failure (Yamaguchi et al., 1976, 1977) and stress dependency (Oda and Koishikawa, 1979; Hettler and Gudehus, 1988; Kusakabe et al., 1991). The measured detailed shear strain distributions are given in Figure 6.9 at different stages of the footing settlement (Yamaguchi et al., 1977). The location of the observed final slip lines detected by radiography is also given in the figure. It is clearly demonstrated that these strains develop along the slip line and at the peak load the magnitudes of the shear strains along the slip line differ considerably for each test series. For the footing of width $B_n = 30\,\text{mm}$, the shear strains in almost all the regions along the slip line remain at around 5%, while the maximum value of shear strain is around 6–7% at the peak. For the footing with $B_n = 1.2\,\text{m}$, the shear strains are generally large and show a considerable variation along the slip line. They concluded that the scale effect can be reasonably explained by progressive failure and the assumption of constant shearing strain adopted in existing bearing capacity theories cannot be valid.

Another explanation is that the dominant reason of the scale effect is the stress dependency of shearing resistance. de Beer (1963) stated that the scale effect is due to the non-linear Mohr–Coulomb's failure envelope. A more fundamental explanation is that the bearing capacity of a footing on sand will be affected by both peak and critical state strengths. Strength (or ϕ') is a function of dilation which would not be uniform beneath a loaded foundation and a complicated calculation is needed if this is to be taken into account. A pragmatic approach is therefore to use a curved failure envelope in a stress characteristic calculation. For example, according to the proposal by de Beer (1963), the stress dependency of ϕ' may be expressed as a semi-log reduction with mean effective stress at failure, as:

$$\phi' = \phi'_o - A \log\left(\frac{\sigma_m}{\sigma_{mo}}\right) \qquad (6.14)$$

where ϕ'_o is the angle of shearing resistance at a reference pressure (σ_{mo}) and A is a parameter indicating the degree of stress dependency. Figure 6.10 demonstrates the results of calculations using stress characteristics to interpret the scale effect (Kusakabe et al., 1992).

Hettler and Gudehus (1988) proposed a method of finding a weighted mean value of ϕ'_m to be used in the bearing capacity formula such as equation (6.8) and calculating N_γ as a function of the footing width, thus taking the scale effect into account from the viewpoint of stress dependency of ϕ'. The proposed method agrees well with the centrifuge test results given by Kimura et al. (1985). This method, however, requires an iterative procedure to determine ϕ'. An alternative way is to use the best straight-line fit to the curved failure envelope in the range of interest without considering the

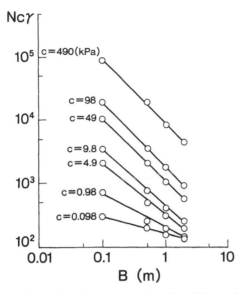

Figure 6.10 Scale effect of combined bearing capacity. $\phi_0 = 40°$, A = 0.138, submerged unit weight $\gamma' = 9806 \, \text{N/m}^3$. (After Kusakabe *et al.*, 1992a. ASCE, with permission.)

variation of ϕ' with stress level (Kutter *et al.*, 1988b). The effectiveness of this approach was supported by four centrifuge tests of circular surface footings of 38.1 and 50.8 mm in diameter on dense dry Montery 0/30 sand ($D_{50} = 0.4 \, \text{mm}$, relative density $D_r = 93\text{–}95\%$), modelling 0.96 m and 1.9 m prototype footings. The stress range to be used is given approximately by $0 < s < 0.6q$, in which s is the mean principal stress and q is the ultimate bearing pressure.

6.2.2.3 Effect of depth of embedment. The magnitude of settlement of footings can be reduced by increasing the depth of embedment of the footings. This effect is often expressed by the ratio of settlement reduction r defined by:

$$r = 1 - \frac{S_d}{S_o} \tag{6.15}$$

where S_o is the settlement of a surface footing, S_d is the settlement of a footing with a depth of embedment D_f. Centrifuge tests have been undertaken on a strip footing by Kimura *et al.* (1985) and Aiban and Znidarcic (1991) and on square footings by Pu and Ko (1988); the results given by Kimura *et al.* (1985) are plotted in Figure 6.11 with respect to the ratio of settlement reduction against $D_f : B$ ratio, together with some theoretical predictions.

6.2.2.4 Effect of anisotropy. A centrifuge study of the effect of anisotropy was made using two kinds of model soils illustrated in Figure 6.12, in which

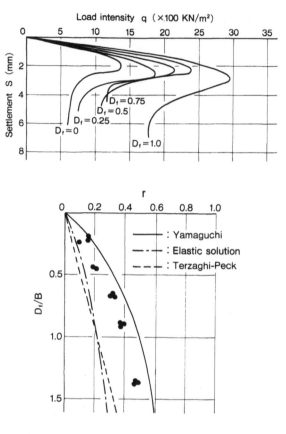

Figure 6.11 Effect of footing embedment. (After Kimura *et al.*, 1985. Thomas Telford, with permission.)

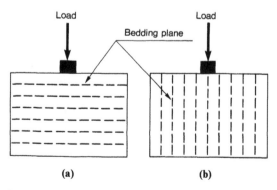

Figure 6.12 Directions of bedding plane in loading tests. (a) V-case, (b) H-case. (After Kimura *et al.*, 1985. Thomas Telford, with permission.)

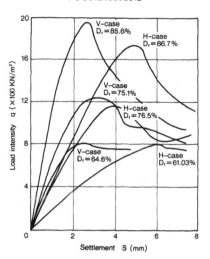

Figure 6.13 Load–deflection curves for V- and H-cases. (After Kimura *et al.*, 1985. Thomas Telford, with permission.)

H-case and V-case refer, respectively, to the bedding plane being parallel and perpendicular to the direction of load application (Kimura *et al.*, 1985).

Loading tests with a strip footing of 30 mm breadth were conducted at 30 *g*. Typical load intensity–settlement curves are presented in Figure 6.13. It is noted that the ultimate bearing capacities for the V-case are slightly larger than those for the H-case, but no difference in the ultimate bearing capacity is apparent for the small value of relative density, D_r. It is interesting to note that the settlements at peak load for the H-case are noticeably larger than those for the V-case regardless of D_r values. The effect of anisotropy is more marked in respect of the settlement behaviour than the bearing capacity.

6.2.2.5 Combined loading.

Eccentricity. Bearing capacity of shallow foundations is reduced considerably by load eccentricity. A widely accepted hypothesis is the concept of effective width as given in equation (6.11).

For the cases where no cohesion and no embedment exist, equations (6.1) and (6.11) give:

$$\frac{Q}{B} = \left(1 - \frac{2e}{B}\right)^2 \gamma \frac{B}{2} N_\gamma \qquad (6.16)$$

The value of $(1 - 2e/B)^2$ is termed the eccentricity factor. Figure 6.14 (Terashi *et al.*, 1984) summarizes the data of the effect of eccentricity including both experiments performed at 1 *g* and on the centrifuge. The concept of the effective width may be accepted as a safe assumption for dry cohesionless

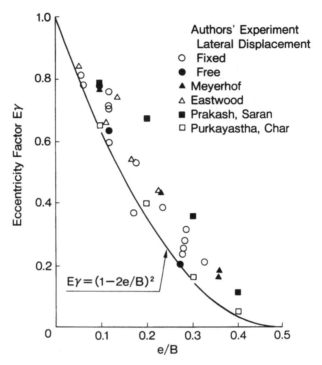

Figure 6.14 Test data of eccentricity factor. (After Terashi *et al.*, 1984. Japanese Society of Soil Mechanics and Foundation Engineering, with permission.)

soils. Aiban and Znidarcic (1991) proposed a slight modification to equation (6.16) based on their centrifuge tests.

General combined loading. The reduction of bearing capacity due to inclined loading is derived from centrifuge tests of strip footings on dense sand (Aiban and Znidarcic, 1991) as:

$$i_r = 1.3 - 3.4 \tan(\alpha) + 3.2 \tan^2(\alpha) \qquad (6.17)$$

The bearing capacity under inclined loading is obtained by multiplying equation (6.9) by i_r.

Yield loci for shallow surface footings were experimentally deduced (James and Shi, 1988; Dean *et al.*, 1993) with special reference to a spud can foundation which is subjected to combined vertical, horizontal and moment loading. Figure 6.15 is a typical interaction locus in V–H space. An empirical equation for describing the limiting yield locus for a particular shape of spud can footing is given by Dean *et al.* (1993).

6.2.2.6 Effects of slopes near footings. The bearing capacity of shallow footings on a slope is a problem which geotechnical engineers frequently

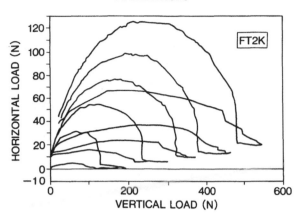

Figure 6.15 Yield locus in V–H plane. (After Dean *et al.*, 1993. Thomas Telford, with permission.)

encounter. The study of the effect of slopes near footings may be applicable to various foundation problems such as conventional spread footings for buildings and bridges, stability of oil storage tank pads, and bearing capacity of backwaters.

Kimura and Saitoh (1981) performed loading tests of surface footings near slopes for various combinations of the following four parameters: the breadth of a footing, B; the height parameter, η; the distance parameter, λ; and the angle of inclination of the slope, β (Figure 6.16). A radiograph is shown in

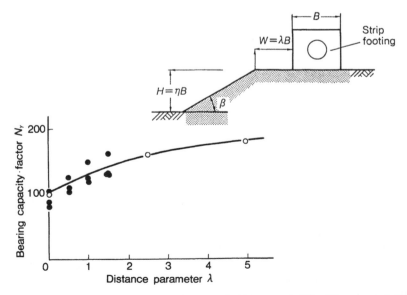

Figure 6.16 Reduction of bearing capacity due to sloping ground. (After Kimura *et al.*, 1985. Thomas Telford, with permission.)

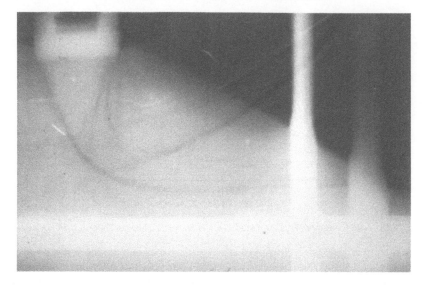

Figure 6.17 Radiograph showing slip surface. (After Kimura *et al.*, 1985. Thomas Telford, with permission.)

Figure 6.17, in which well-developed failure lines can be seen. The failure line of the passive zone is almost horizontal, like the failure mechanism predicted by the upper bound calculation of Kusakabe *et al.* (1981). A relationship between the bearing capacity factor N_y and the distance parameter obtained from the centrifuge tests is given in Figure 6.16. The solid points are the results of large-scale model tests reported by Shields *et al.* (1977) for silica sand under similar experimental conditions. Although the angles of shearing resistance are slightly different, the agreement between the two is remarkable.

Gemperline and Ko (1984) conducted 33 similar loading tests of a strip footing on a dry slope. They compared the maximum bearing pressure with numerical solutions by using seven analytical methods then available. Gemperline (1988) continued this research work and carried out a more sophisticated testing programme, investigating geometrical variables that included footing width (B), footing depth to width ratio ($D:B$), slope angle, the ratio of distance between the slope crest and the footing edge to the width of the footing ($b:B$). Four different sands were used. A total of 215 tests were performed, representing 194 combinations of test variables. From this comprehensive test programme, he presented experimentally deduced bearing capacity formulae for a shallow footing at the top of a cohesionless slope. This is a clear demonstration of the power of centrifuge modelling to examine complex three-dimensional foundation problems.

6.2.2.7 Some modelling considerations.

Footing roughness. Conditions at the base of footing a can be assumed to be perfectly rough in most practical situations. It is common practice to ensure a rough condition by gluing sand particles of the model soil over the base of the footing. Failure to achieve a perfectly rough condition might influence the behaviour of a footing, as demonstrated by Kimura *et al.* (1985).

Loading system. In the cooperative study on the bearing capacity of footings on a saturated soil, reported by Corté *et al.* (1988), the three different organizations used their own loading system; two were stress-controlled and one strain-controlled. Figure 6.5 showed the load intensity–settlement curves from each organization. It is obvious that there exists a slight scatter, in particular in the value of sub-grade reaction.

Kitazume (1984) demonstrated the importance of freedom of footing rotation by conducting concentric and eccentric loading tests together with finite-element analysis. It was shown that inadequate simplification of this detail may lead to completely different results in centrifuge models.

Another example of the influence of fixity at the loading point is seen in Figure 6.18 (Kutter *et al.*, 1988b), in which a sharp drop is observed after the maximum bearing pressure in the case of a footing which is free to rotate,

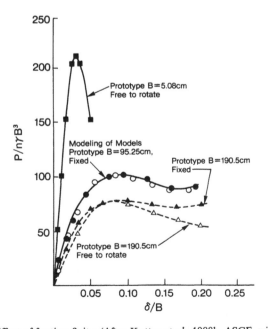

Figure 6.18 Effect of footing fixity. (After Kutter *et al.*, 1988b. ASCE, with permission.)

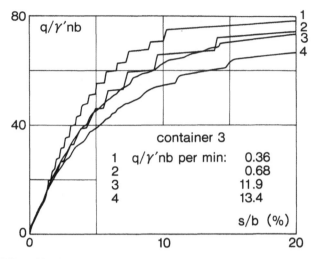

Figure 6.19 Effect of loading rate on load–deflection curves. (After Corté *et al.*, 1988. Balkema, with permission.)

whereas the bearing pressure re-increases somewhat after the maximum bearing pressure in the case of a footing being fixed.

Rate of loading.   Selection of appropriate loading rates is generally not a crucial decision for dry sand, but may become important when the sand is saturated. Figure 6.19 shows influence of loading rate of a circular footing on saturated sand (Corté *et al.*, 1988), in which it is seen that the lower the loading rate, the higher the soil resistance. This may be associated with possible development of excess pore-water pressures and measurement of pore-water pressure is advisable when using saturated cohesionless soils.

Reproducibility.   Sources of experimental error stem from the preparation of the model, acceleration history, deviation of acceleration from the pre-scribed value, boundary conditions such as the loading system, base rough-ness and freedom of movement of the footing. It is important to assess the general magnitude of error in centrifuge loading tests. Gemperline (1988) performed 26 tests representing the same prototype of a footing resting near the edge of a cohesionless slope, from which he obtained a coefficient of variation of 9.5%.

Grain-size effect.   It is now well known that shear rupture propagation is affected by particle size. The influence of grain size of soil on bearing capacity has been examined by several research workers. Yamaguchi *et al.* (1977) used glass ballotini of six different gradings to give a footing width to mean grain diameter ratio ($B:D_{50}$) of 307–55. Figure 6.20 presents the load intensity–settlement curves from which they concluded that there was little influence on

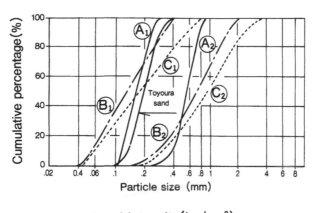

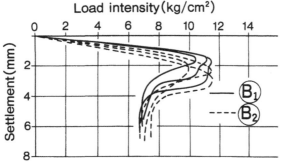

Figure 6.20 Effect of grain size on load–deflection curves: grading curves of glass ballotini and Toyoura sand. (After Yamaguchi *et al.*, 1977. Japanese Society of Soil Mechanics and Foundation Engineering, with permission.)

bearing capacity. Gemperline (1988) also examined the footing–grain-size effect by using two different footing widths, 25.4 mm and 50.8 mm, for a sand having average grain sizes of 0.60 mm, giving a foundation width to grain-size ratio of 84–42. He observed a distinct difference in the behaviour of two out of three 25.4 mm tests near the maximum bearing pressure. Bolton and Lau (1988) conducted a careful examination on the scale effect arising from particle size. They used two silica sands which had particle gradings which differed by a factor of 50 in respect of particle size. They pointed out the important issue relating to scaling on particle size, which is particle crushing; the smaller the particles, the larger the crushing pressure.

Tatsuoka *et al.* (1991) reported the comparison of footing tests on dry Toyoura sand ($B:D_{50}$ ratio of 150) as shown in Figure 6.21, in which centrifuge tests yield larger maximum bearing pressures and larger settlements at the maximum pressure, compared to corresponding 1 *g* model tests. They argued that the shear band width was proportional to the particle size. The difference between the centrifuge results and 1 *g* tests was caused by a particle-size effect. Siddiquee *et al.* (1992) demonstrated the capacity of their

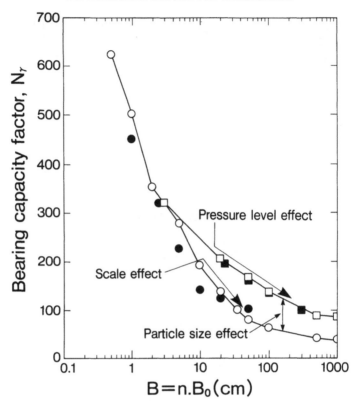

Figure 6.21 Comparison of centrifuge tests and $1\,g$ tests. $N_\gamma = \left(\dfrac{2q}{\gamma B}\right)_{max}$. $\square$ Centrifuge test results

($B_0 = 3\,$cm), $\square$ centrifuge simulation, $\bullet$ $1\,g$ test results, $\bigcirc$ $1\,g$ simulation. (After Siddiquee *et al.*, 1992. Japanese Society of Soil Mechanics and Foundation Engineering, with permission.)

constitutive model to explain the difference, and the results are also presented in Figure 6.21.

From this evidence, it is clear that there exists to some extent a grain size effect if the foundation width to grain-size ratio is less than about 50–100. However, this does not mean that the centrifuge models are of no value. One of the simple solutions is to use a larger footing, giving an acceptable size ratio, say $B:D_{50}$ greater than 100. It must be remembered, as Bolton and Lau (1988) emphasized, that if particle scaling is carried out for the purposes of satisfying scaling problems related to rupture formation or permeability, it must be appreciated that the relative resistance to crushing of smaller particles may influence the results.

Undistributed samples. It is important to preserve *in situ* fabrics of granular materials in an attempt to model a real foundation problem. For example, the

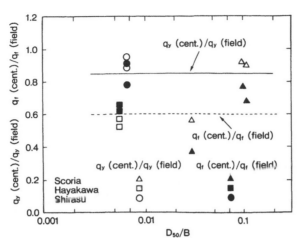

Figure 6.22 Comparison of ultimate and yield loads between field tests and centrifuge tests using undisturbed soils. (After Kusakabe, 1993.)

deterioration of real cohesion would lead to underprediction of bearing capacity. Three cases of direct comparison between the centrifuge tests and *in situ* large-scale loading tests on natural granular materials having $B:D_{50}$ ratios of 100–150 (Shirasu), 9–32 (Scoria) and 179 (Hayakawa) are collected and the data are summarized in Figure 6.22. Note here that except in the case of Shirasu, the block samples experienced the process of freezing and thawing before the centrifuge loading tests (for details of the experiments using Scoria, see Kusakabe *et al.*, 1992b). It can be seen that centrifuge tests underpredict yield bearing capacity by 15%, and ultimate bearing capacity by 35%; in both cases, the predictions are on the safe side. However, settlements in the models at yield loads are always larger than those of field tests as seen in Figure 6.23. Consequently, initial stiffnesses (sub-grade reaction) in models are less than those of the prototype, although stiffnesses during unloading and reloading are of the same order of magnitude compared to the field test. These discrepancies may be due to the combined effect of sample disturbance, and grain-size effect and bedding error. Further studies are needed to clarify the differences.

6.2.3 Shallow foundations on cohesive soils

Bearing capacity problems of shallow foundations on fine-grained soils are generally concentrated on the short-term undrained behaviour. Subsequent consolidation behaviour is also interesting under working load conditions. Centrifuge model tests of shallow foundations on fine-grained (clay) soils are relatively more complicated than those using coarse-grained (sand) soils. The first difficulty is in achieving a predetermined strength profile in the model.

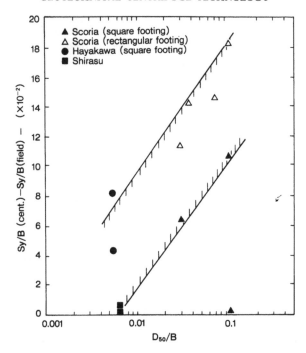

Figure 6.23 Data of the difference of settlement in field and centrifuge models plotted against D_{50}/B. (After Kusakabe, 1993.)

The second difficulty is that of lengthy consolidation times. The third difficulty is the instrumentation. Measurements of pore-water pressure are essential if undrained behaviour is to be interpreted in terms of effective stress.

When a layer of clay is weak, some type of soil improvement is generally used in order to increase the stability of foundations. The effectiveness of soil improvement methods may also be examined by centrifuge modelling.

6.2.3.1 Normally consolidated clay.   For a homogeneous deposit, there is a decrease in void ratio with depth due to consolidation by self-weight of the soil. This decrease in void ratio results in an increase in undrained shear strength c_u with depth, especially for normally consolidated clays. Apart from the upper layer of desiccated clay, the change of strength with depth is approximately linear. The effect of the increase with depth is negligible when the footings are small, but for larger footings, the effect can be substantial and serious consideration should be given to the variation of strength with depth. Centrifuge modelling is suitable for this purpose.

The bearing capacity formulae for foundations on a normally consolidated clay layer are given by equations (6.5) or (6.6). An experimental verification of the two-dimensional theoretical values of N_{co} was attempted by Nakase *et al.* (1984). The parameter of kB/c_o was found to lie in the range of 0.03 to 4.58 for

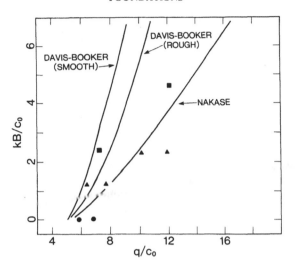

Figure 6.24 Comparison of bearing capacity factor between theories and centrifuge tests. ● 1 g, ▲ 40 g, ■ 80 g. (After Nakase *et al.*, 1984. ASCE, with permission.)

various combinations of pre-consolidation stress (and therefore c_0 values) and level of centrifuge acceleration. The results of centrifuge tests compared with the analytical results using the slip circle method by Nakase (1966) and plasticity solutions (Davis and Booker, 1973) are shown in Figure 6.24. Both theoretical methods make use of a rigid-plastic soil model. The measured values of bearing capacity are located around the slip circle solution and the plasticity solution, which gives confidence in using equations (6.5) and (6.6) with the bearing capacity factor of N_{co} adopted in current practice.

6.2.3.2 Multi-layer of cohesive soils. If a circular caisson is to be built on a thin layer of clay overlying a stiff layer, an expression similar to equation (6.7) may be used to estimate bearing capacity. Kutter *et al.* (1988a) examined this type of problem. The model geometry consisted of a layer of clay sandwiched by upper and lower dense sand layers. The clay layer was slightly over-consolidated. A plastic cylinder with a flexible membrane base was located on the upper sand layer and was filled with water after pore-pressure equilibrium in the clay was achieved. At failure, the bearing capacity for the circular footing was found to be given by:

$$q = c_u \left(5.5 + \frac{2r_0}{3t} \right) \tag{6.18}$$

in which r_0 is the radius of caisson and t is the thickness of clay layer.

The ratios of predicted to observed bearing capacities were in the range 1.03–1.70. The c_u value used in equation (6.18) was based on the average value of c_u at the mid-depth beneath the edge and at the mid-depth beneath the centre of the caisson evaluated from triaxial compression test results.

Kusakabe (1980) carried out a limited test series with a similar loading condition and model geometry, consisting of an upper loose sand layer overlying a normally consolidated (NC), and an overconsolidated (OC), clay layer. The failure surface was observed to pass through the boundary between the NC and OC layer and a conventional two-dimensional slip surface seemed to be acceptable by using the c_u value at mid-depth of the NC layer.

The existence of a hard surface crust leads to a substantial increase in the bearing capacity and the initial rigidity. Nakase et al. (1987) developed a technique of making a hard crust and performed a series of two-dimensional loading tests on various strength profiles of clay layer as illustrated in Figure 6.25. It was found that as the strength of the surface crust increases, the bearing capacity increases and deformations extend to greater depths. A simple failure mechanism which gives reasonable predictions of the bearing capacity was found.

6.2.3.3 Some aspects about modelling.   Selection of rate of loading is always associated with drainage conditions in cases of cohesive soils. Production of perfectly undrained conditions may be difficult, in particular if small size models are to be tested under high g. Drainage conditions at the base of a foundation may need special attention for larger-scale models. It is always helpful to have pore-pressure measurements while performing tests of shallow foundations on fine-grained soils, to allow interpretation of phenomena in terms of effective stress.

6.3 Pile foundations

There exist a number of types of pile foundations, which may be classified in different ways such as type of materials (e.g. steel, concrete and timber), and type of manufacture (pre-cast, cast-in-place), type of pile shape (e.g. pile, H-pile, taper, open-ended or closed-ended), and type of installation methods (e.g. driven, bored). It is well known that the behaviour of a pile is significantly affected by installation methods. Also, it is generally known that the responses of a pile under vertical loading are more sensitive to the installation method, compared with those under horizontal loading.

6.3.1   Axial loading

6.3.1.1   Effect of installation.   It is now well known that pile installation should be performed in flight in order to establish a stress regime around the pile which is comparable to that of a field prototype. The effect of lack of

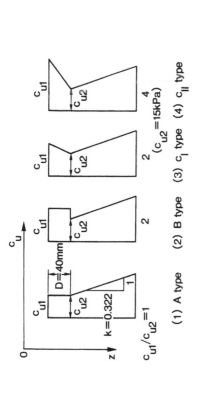

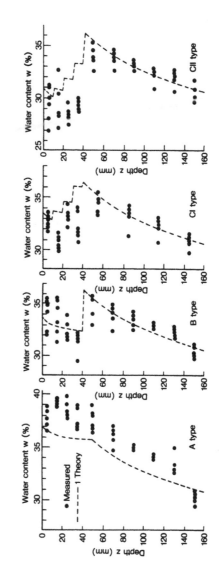

Figure 6.25 Various types of strength profile produced in models. (After Nakase *et al.*, 1987. Japanese Society of Soil Mechanics and Foundation Engineering, with permission.)

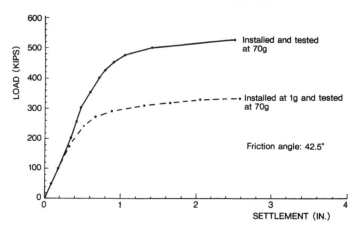

Figure 6.26 Effect of pile installation at different *g* levels. (After Ko *et al.*, 1984. ASCE, with permission.)

similarity of installation procedure can lead to serious errors in predictions of axial capacity. Figure 6.26 shows clear evidence of the effect of in-flight installation (jacked in place at 70 *g*) versus 1 *g* installation on a single model pile in dense sand (Ko *et al.*, 1984). The bearing capacity of the pile installed at 1 *g* is about 60% of that of the pile installed at 70 *g*. The effect of installation acceleration level was examined in detail by Craig (1984). He pointed out that reductions in capacity of up to 50% can result from 1 *g* installation. Results of tests using 1 *g* installation should be interpreted with care.

6.3.1.2 Modelling of models. Ko *et al.* (1984) performed the modelling of models on axial bearing capacity of a single pile jacked in flight into dense

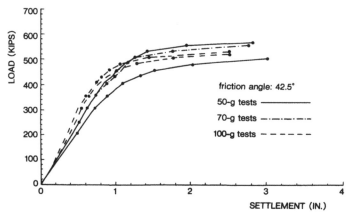

Figure 6.27 Load–settlement curves for modelling of models. (After Ko *et al.*, 1984. ASCE, with permission.)

sand, using three model piles fabricated at 1/50, 1/70 and 1/100 scale. As is seen in Figure 6.27, similar load–settlement curves were obtained, verifying the similitude of scaling, although the data of the $50\,g$ tests show some deviations from other results. Nunez *et al.* (1988a) report an attempt of modelling of models on a single pile installed in dense sand, using a driving hammer in-flight, and then subjected to tension loading. In the tests, consistently higher capacities were measured for smaller piles tested at the higher g level. The differences in capacity were about 30%. The reasons for the disparity were considered to be differences of the pile wall thickness and the ratio of the mass of hammer piston used to drive the piles to the mass of the pile.

6.3.1.3 Bearing capacity factor. The axial capacity of a pile comes from the base resistance and shaft resistance. The values of bearing capacity factor, N_q, and of skin friction are of practical interest. Figure 6.28 shows the variation of N_q value with the overburden effective stress, indicating a decrease with an increase of overburden pressure. In the range of $\sigma'_v = 50 - 500\,\mathrm{kN/m^2}$, the value of N_q is reduced by a factor of 2. This may be called the stress level effect of N_q, which may be compared with the scale effect of N_γ shown in Figure 6.7. Both effects may have the same explanation, that of the variation in mobilized angle of friction with an increase in mean stress level.

An example of the peak skin friction profile is given in Figure 6.29 (Nunez *et al.*, 1988a), showing the linear increase with depth. By knowing the vertical stress gradient at the centrifuge test acceleration, the stress ratios of skin friction to effective overburden, known as β values, were $\beta = 1.51$–2.2 which is consistent with the current design chart (Poulos and Davis, 1980). In the tests,

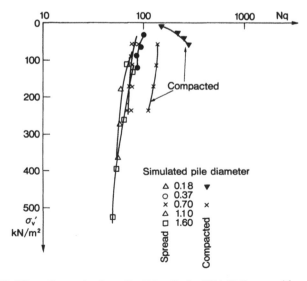

Figure 6.28 Effect of stress level on N_q. (After Craig, 1984. Balkema, with permission.)

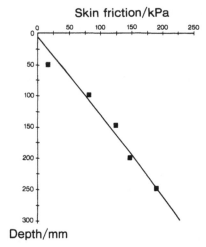

Figure 6.29 Variation of skin friction with depth. (After Nunez *et al.*, 1988a. Balkema, with permission.)

there was no evidence of a critical depth below which the skin friction reached a limiting value.

For fine-grained soils, Ko *et al.* (1984) reported that a pile base takes from 5 to 15% of the applied load with the remainder of the load transferred fairly uniformly along the pile length. The ratio of unit shaft resistance to undrained shear strength, known as the α value, may be deduced from the test data quoted by Ko *et al.*

6.3.1.4 Full-scale field tests. Ko *et al.* (1984) reported the direct comparison between a full-scale field test and corresponding centrifuge test on a

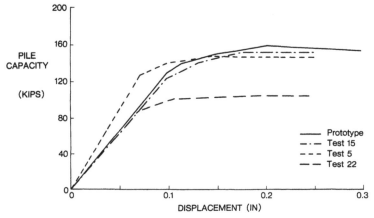

Figure 6.30 Comparison of load–settlement curves between prototype and centrifuge tests. (After Ko *et al.*, 1984. ASCE, with permission.)

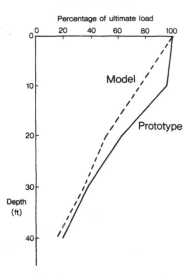

Figure 6.31 Load transfer comparison between prototype and centrifuge tests. (After Ko *et al.*, 1984. ASCE, with permission.)

single pile driven into a clay stratum subjected to monotonic axial loading. The strength profile of clay layer had a slight linear increase with depth with a c_u of 115 kPa at the mid-depth of the pile. The actual value measured by in-flight vane ranged from 91 to 139 kPa. Comparison of load–deflection curves (Figure 6.30), load transfer curves (Figure 6.31) and pile capacity against undrained shear strength indicates that centrifuge data are generally in good agreement with the full-scale field tests.

An attempt was also made by the same authors to simulate a full-scale field test behaviour of a pile in sand, but the multi-layer system of the prototype soil profile hampered direct comparison between the model and full-scale field test.

6.3.1.5 Cyclic loading. A number of studies on piles subjected to axial cyclic loading have been performed related to the foundation of offshore structures. The main concern is the degradation of base and shaft resistance due to cyclic wave loading. The magnitude of cyclic load at failure is controlled by the level of the applied static load and the size of load increment and total number of cycles. Both piles installed in sand and clay were examined in centrifuge tests.

Sabagh (1984) performed a series of cyclic loading tests on a single pile installed in-flight. Figure 6.32 illustrates a loading sequence from the beginning of penetration to subsequent cyclic loading. Failure was defined according to one of the following criteria. First a ratio of cyclic displacement, δ_v, to pile diameter, d, exceeding 0.01. Secondly, the cumulative pile top displace-

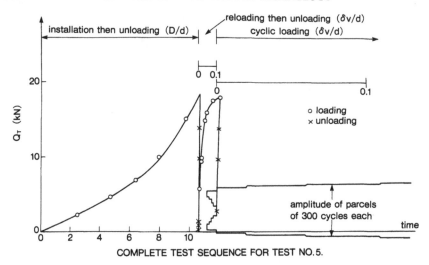

Figure 6.32 Loading sequence for cyclic loading. (After Sabagh, 1984. Balkema, with permission.)

ment exceeding $0.1d$. Loading tests with a cyclic load, Q_T^c, were performed with various ratios of total static pile load, Q_T^s, to total failure axial load, Q_T^f and obtained the failure envelope shown in Figure 6.33.

When the ratio of $Q_T^s : Q_T^f$ was greater than 0.26, the piles failed in compression associated with the large cumulative movement. Repeated cyclic loading reduced both shaft and base capacity relative to the static condition, with the base load appearing most sensitive. In contrast, the piles failed in tension with large cyclic movements and the pile displaced upwards in the case of $Q_T^s : Q_T^f$ less than 0.11. Cyclic loading reduced shaft capacity relative to the static case.

A similar test series of cyclic loading test in dense sand was carried out by Nunez *et al.* (1988a), who reported the distribution of load transfer profiles under compression and tension. They also examined the effect of re-driving.

6.3.1.6 Pile groups. The influence of the order of driving piles is well known in practice, in particular for piles in sands and gravels. Figure 6.34 shows an example of driving resistances of eight piles driven into dense sand in flight, and the driving order is indicated in the figure (Ko *et al.*, 1984). Again, simulation of driving order at correct stress level becomes a very important aspect of behaviour of pile groups. Subsequent axial loading tests showed a group efficiency of 1.15–1.32 for dense sand and 0.96–1.04 for clay.

6.3.2 Lateral loading

6.3.2.1 Governing equation. It is known that the lateral response of piles is largely independent of pile length provided the slenderness ratio is above a

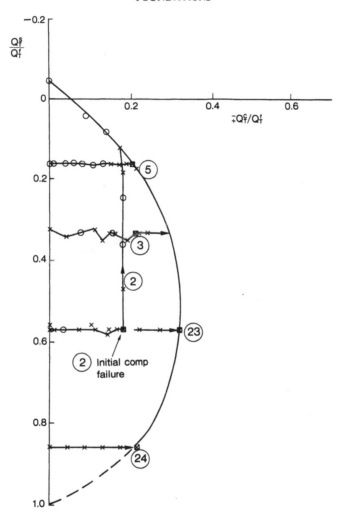

Figure 6.33 Yield surface for cyclic loading. Group 1 smooth pile (35 mm diameter). ○ points corresponding to a net upward movement over the parcel of cyclic loading, × points corresponding to a net downward movement over the parcel of cyclic loading, □ failure point. Circled numbers indicate test numbers. (After Sabagh, 1984. Balkema, with permission.)

critical value. In common practical design of a laterally loaded pile, a sub-grade reaction analysis is often adopted, where the pile is assumed to act as a thin strip whose behaviour is governed by the beam equation:

$$E_p I_p \frac{\mathrm{d}^4 y}{\mathrm{d}x^4} = -pd \qquad (6.19)$$

where E_p is the modulus of elasticity of the pile, I_p is the second moment of

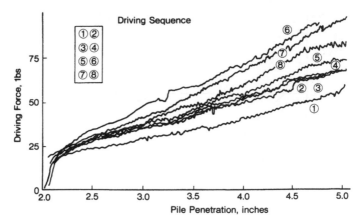

Figure 6.34 Effect of driving sequence. The first 2.0 inches of pile were pushed into the soil at 1 g. Friction angle 41.35°. (After Ko *et al.*, 1984. ASCE, with permission.)

area of pile section, p is the pressure from soil, x is the depth in soil, y is the deflection of the pile at depth and d is the width or diameter of pile.

If a relationship between the soil pressure and the deflection of a pile (p–y relation) is known, equation (6.19) can be solved. p–y curves may be studied experimentally by conducting lateral loading tests on a long pile with measurements of strain distribution along the pile length at different lateral pressures. The following equations are used:

$$y = \iint \frac{M}{EI} \tag{6.20}$$

$$p = \frac{\mathrm{d}^2 M}{\mathrm{d}x^2} \tag{6.21}$$

where M is the bending moment.

6.3.2.2 Effect of installation. It was previously pointed out that in-flight installation of a pile is an essential requirement when studying axial capacity. In the cases of laterally loaded piles, the effect of pile installation at a lower stress level is generally thought to be less pronounced. Figure 6.35 shows a comparison between lateral load–lateral deflection curves for piles installed at 52.5 g and at 1 g (Craig, 1984). Although the initial stiffness of a pile installed at 1 g is smaller than that of a pile installed at 52.5 g (by about 10% for a stiffer pile), the effect on the overall behaviour is less significant.

Various installation methods including jacking, hammering, placement in boreholes or pluviation of sand around the pile (presumably at 1 g) were tried and no significant effect was reported (Bouafia and Garnier, 1991).

Comparison with full-scale field test data provides further evidence for this

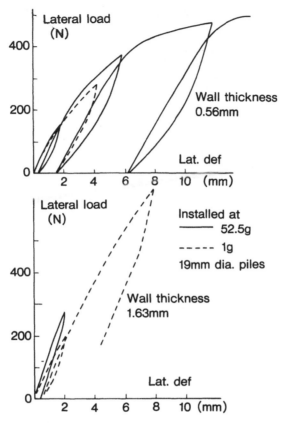

Figure 6.35 Effect of installation *g* level on lateral loading tests. (After Craig, 1984. Balkema, with permission.)

conclusion. Thus the majority of the following data presented was obtained by piles installed at 1 *g*, except the work by Oldham (1984) who drove piles in flight and subsequently applied lateral loads.

6.3.2.3 Modelling of models. Modelling of models on a single plate pile in dense sand subjected to lateral loading using a model test arrangement shown in Figure 6.36 is reported by Terashi *et al.* (1989). Six tests were performed at different accelerations of 25.7–77.2 *g*. There were slight differences between the six model piles in terms of pile width and pile rigidity (*EI*). Figure 6.37 presents the data of six test results of lateral load plotted against lateral displacement at the pile top at the corresponding prototype scale. Figure 6.38 shows the bending moment distribution along the pile length for the same test series. From these two figures, it is clear that the prototype behaviour may be examined using centrifuge modelling.

Two other modelling of model tests were performed on a single pile

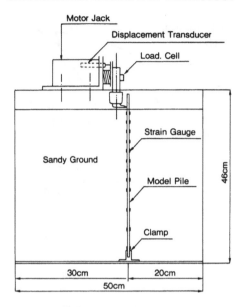

Figure 6.36 Typical set-up for laterally loaded pile tests. (After Terashi *et al.*, 1989.)

installed into a quartz sand at 1 *g* without surcharge pressure (Barton, 1984) and a calcareous sand at 1 *g* with a surcharge pressure subjected to cyclic loading (Nunez *et al.*, 1988b). The result of Nunez's tests on load–deflection curves of two tests at 52 *g* and 92 *g* is shown in Figure 6.39. The results show a high degree of similarity and validate the modelling procedures.

Modelling of models was also performed on piles in normally consolidated

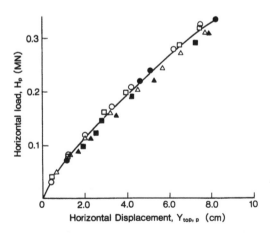

Figure 6.37 Modelling of models for laterally loaded piles in terms of horizontal displacement and horizontal load. ○ 25.7 *g*, △ 28.6 *g*, □ 40.2 *g*, ● 49.2 *g*, ▲ 66.4 *g*, ■ 77.2 *g*. (After Terashi *et al.*, 1989.)

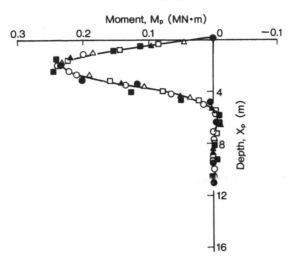

Figure 6.38 Modelling of models for laterally loaded piles in terms of moment distribution with depth. ○ 25.7 g, △ 28.6 g, □ 40.2 g, ● 49.2 g, ▲ 66.4 g, ■ 77.2 g. (After Terashi *et al.*, 1989.)

clay (Hamilton *et al.*, 1991). A test was conducted at 46 g on a model pile that was 27 mm diameter by 230 mm embedded length. Another test was conducted at 93.4 g on a model pile that was 13.3 mm diameter by 1150 mm embedded length, both simulating the prototype event of a pile of 1.24 m

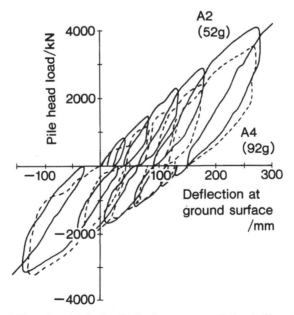

Figure 6.39 Modelling of models for load deflection curves at pile head of laterally loaded piles. (After Nunez *et al.*, 1988b. Balkema, with permission.)

diameter and 10.58 m embedded length. Fairly good agreement was obtained
between the two.

6.3.2.4 p–y curves.   Load–deflection or *p–y* curves are obtained from the
bending moment distribution, by fitting a curve to the data which are then
differentiated twice with respect to depth to give the soil resistance and
integrated twice to obtain the deflection of the pile according to equations
(6.20) and (6.21). The integration is a very stable procedure and reliable
estimates of deflection are easy to obtain. The double differentiation is very
dependent on minor details of the moment distribution curve. Previous
research workers used either polynomial functions (Lyndon and Pearson,
1988; Terashi *et al.*, 1991), or spline functions (Oldham, 1984; Finn *et al.*, 1984;
Bouafia and Garnier, 1991) or a combination of the two (Scott, 1981). The
measured displacement and rotation at the pile head are usually used as the
constants of integration. A typical *p–y* curve is shown for sand in Figure 6.40
(Scott, 1981); results for tests in clay are presented by Hamilton *et al.* (1991).

The *p–y* relationship at a particular depth may be plotted on a logarithmic
scale as in Figure 6.41. From the figure it can be seen that the soil resistance is
proportional to the square root of the deflection. Figure 6.42 also presents the
data of soil resistance versus depth indicating a linear increase of soil
resistance with depth. By combining the results in Figures 6.41 and 6.42, an
experimentally deduced *p–y* relation can be written as:

$$p = k_s x y^{0.5} \qquad\qquad (6.22)$$

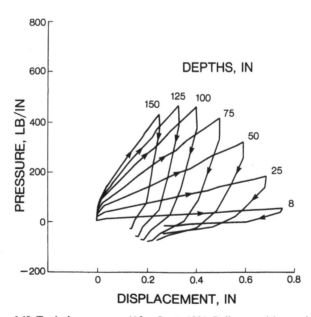

Figure 6.40 Typical *p–y* curves. (After Scott, 1981. Balkema, with permission.)

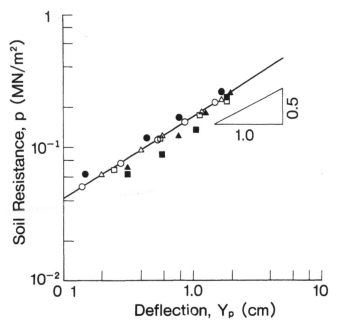

Figure 6.41 Relationship between soil resistance and deflection for various modelling scales. ○ 25.7 g, △ 28.6 g, □ 40.2 g, ● 49.2 g, ▲ 66.4 g, ■ 77.2 g. (After Terashi *et al.*, 1989.)

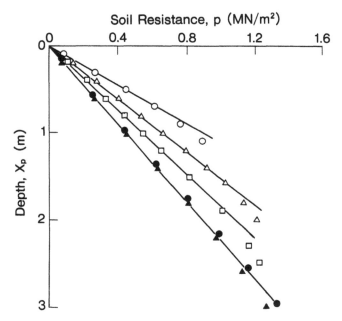

Figure 6.42 Linear variation of soil resistance with depth. EI_p (MN m²): △ 5.11 × 10², ● 1.65 × 10², □ 39.7, △ 14.6, ○ 2.61. (After Terashi *et al.*, 1989.)

which is identical to that derived from large scale model tests for sandy ground by Kubo (1965).

A study of the influence of pile width on the p–y relationship indicates that equation (6.22) still holds with the value of k_s decreasing with an increase of pile width and approaching a constant value when the prototype pile width exceeds about 80 cm.

Hamilton et al. (1991) show that results of p–y curves obtained from the centrifuge model tests on a single pile in normally consolidated clay agree well with the analytical method proposed by Matlock (1970).

6.3.2.5 Effects of pile roughness and installation in slopes. Wall friction is an important parameter in the study of pressure on retaining structures. Similarly wall friction may influence the value of lateral resistance of a pile, although practical design usually does not take into account this effect. Lyndon and Pearson (1988) conducted a series of tests investigating the effect of skin friction of piles in dry dense sand. Figure 6.43 demonstrates the influence of surface roughness on the load–deflection curve and bending moment distributions, indicating that the roughness of the pile surface reduces the deflection by about 40%, and increases the maximum bending moment by about 30%. A similar observation was made by Bouafia and Garnier (1991) who reported that the reduction was 15%.

The influence of sloping ground on lateral resistance was examined in dry sand by Terashi et al. (1991). The result reveals that equation (6.22) does hold even for piles in a slope, by assuming the ground to be horizontal. The influence of the slope appears at a distance somewhere between 2.5B and 1B from the crest where B is the width of the plate pile.

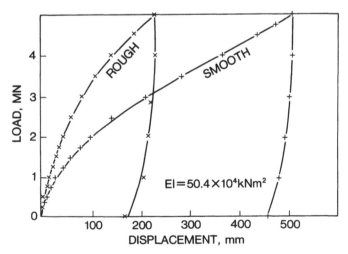

Figure 6.43 Influence of pile surface roughness on load–deflection curves. (After Lyndon and Pearson, 1988. Balkema, with permission.)

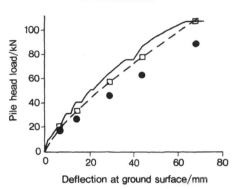

Figure 6.44 Comparison of load–deflection curves between full-scale test and centrifuge tests. Solid line, pit test; dashed line, analysis from Wesselink *et al.* (1988); ● measured A1 □ deduced A1, from centrifuge tests. (After Nunez *et al.*, 1988b. Balkema, with permission.)

6.3.2.6 Cyclic loading. Some research has been undertaken on cyclic lateral loading tests (Scott, 1981; Hoadley *et al.*, 1981; Oldham, 1984). Cyclic loading was not found to have any significant softening effect on the response of the pile–soil system, although a slight stiffening of the soil has been observed (Hoadley *et al.*, 1981).

6.3.2.7 Full-scale test. A set of data is available for direct comparison between full-scale tests and centrifuge tests (Nunez *et al.*, 1988b; Wesselink *et al.*, 1988). Two laterally loaded pile tests were carried out in an onshore pit filled with uncemented calcareous sand. The onshore pit pile had a diameter of 0.356 m, a bending stiffness of 24.0 MN m^2, a pile length of 6.0 m and was driven in the pit. The corresponding centrifuge model piles were driven at 1 g under a surcharge pressure corresponding to an average vertical effective stress at a depth of 3 m and tested at 17 g. At prototype scale, the piles had a diameter of 0.356 m, a bending stiffness of 14.4 MN m^2 and a pile length of 6.45 m. Direct comparison was made of the pile head load–deflection curve (Figure 6.44) and bending moment distribution (Figure 6.45). Very good agreement was observed between data measured and deduced (for corresponding pile stiffness) from the centrifuge test and the values obtained for a corresponding pile from the pit tests. This close agreement implies that the effect of grain size did not influence the pile behaviour significantly.

6.3.2.8 Moment capacity of short pile. The ultimate moment capacity of short piles in cohesionless soil was examined theoretically by Broms (1964), who assumed a passive resistance equal to three times the Rankine pressure in front of the pile and neglected any active pressure behind it.

Verification of the theoretical prediction was offered by Dickin and Wei (1991) who performed centrifuge tests of short piles embedded in dry sand for dense (peak friction angle of 46–49°) and loose (peak friction angle 37–40°)

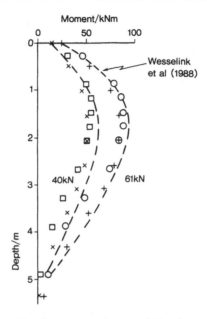

Figure 6.45 Comparison of bending moment between full-scale test and centrifuge tests. □, ○ Pit test; ×, + centrifuge test A1; dashed line, analysis. (After Nunez *et al.*, 1988b. Balkema, with permission.)

soils. Figure 6.46 shows that comparison between moment limits from centrifuge tests and predictions by the Broms formula. The Broms formula gives reasonable agreement for piles in dense sand but yields considerably higher values for those in loose sand.

6.3.2.9 Pile groups. The response of a pile group to lateral loading was examined by Barton (1984) and Nunez *et al.* (1988b), and interaction factors were determined experimentally and compared with predictions for piles in an elastic continuum. Barton deduced that the predictions assuming elasticity underpredicted pile group interaction for closely spaced piles, whereas at larger spacing the interaction factors were overpredicted. The observation of Nunez *et al.* was that the agreement between experimentally deduced and predicted interaction factors was reasonable. These apparently inconsistent observations may have stemmed from the different methods of pile installation.

6.3.3 Some modelling considerations

6.3.3.1 Pile dimensions and materials. Dimensions of model piles are determined by many factors. Model piles are scaled down basically by a factor of N, if the same materials are used. Alternative dimensions may be used in order

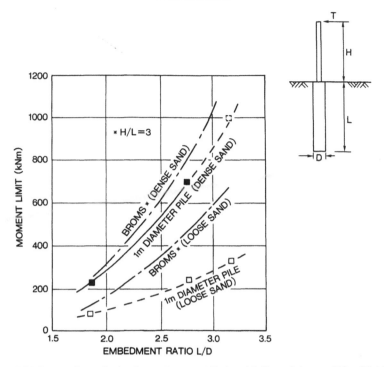

Figure 6.46 Comparison of experimental moment limits with Broms' theory. (After Dickin and Wei, 1991. Balkema, with permission.)

to satisfy some non-dimensional parameters which govern the phenomena to be examined.

Nunez and Randolph (1984) discuss the similarity requirement for tension piles and make use of the flexibility ratio defined by $2(l/d)G/E_p$ (where l is the pile length, d is the pile diameter, E_p is the Young's modulus of an equivalent solid pile and G is the shear modulus of the soil), since the relative stiffness between pile and soil is the significant similarity requirement for cyclic loading of the piles.

Pile dimensions are considered relative to particle size. Generally speaking, it is recommended that the model pile diameter should be maximized under other experimental restraints such as container size, loading capacity and method of installation. Reported examples of ratios of pile circumference to particle size are 150–600 (Nunez et al., 1988a), 360 (Lyndon and Pearson, 1988) and 220–610 (Dickin and Wei, 1991).

No significant particle size effect has been reported for lateral loading conditions. There might be some effect of the ratio of pile circumference to particle size when considering plugging of open ended piles. The wall thickness:diameter ratio may also influence the formation of plugging, resulting in some scale effect on the behaviour of axially loaded open-ended piles.

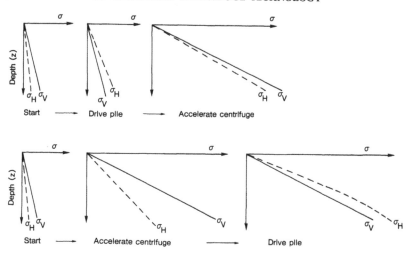

Figure 6.47 Influence of pile installation g level on stress state before loading. (After Craig, 1984. Balkema, with permission.)

6.3.3.2 Method of installation.   It has previously been seen that the effect of installation method on pile behaviour is pronounced in axial loading conditions and less pronounced in lateral loading conditions. Figure 6.47 illustrates the effect of test sequence on stress distribution along the centre-line of a model pile and explains why pile installation should be performed in flight (Craig, 1984). Pile installation involves some soil displacement, which induces stress changes in the ground. At prototype scale, changes occur in the high mean normal stress with less dilation than in the corresponding model pile installed at 1 g in dense sands where the lower stress levels would lead to greater dilation and a higher initial earth-pressure coefficient around the pile. Installation at higher stress would cause some particle crushing during installation.

Craig (1984) further pointed out that the method of installation is interrelated with the selection of pile dimensions. Limitations of working conditions at maximum acceleration levels must be considered in respect of: (i) installation buckling restraint, (ii) boundary effect restraints; and (iii) acceleration restraints.

A group of centrifuge workers at UMIST developed installation systems which could 'push-in' piles during centrifuge flight. Basic components of the system were a pneumatic jack and electric control units which operated pneumatic valves and regulators to control the push-in of a vertical pile and subsequent axial cyclic loading (Cook and Lewis, 1980). This system was later modified to incorporate a lateral loading capability (Craig, 1984). A similar system of push-in installation was used at the University of Colorado, in which a double acting bellofram cylinder was utilized (Ko *et al.*, 1984). An alternative system of a push-in installation system is shown in Figure 6.48

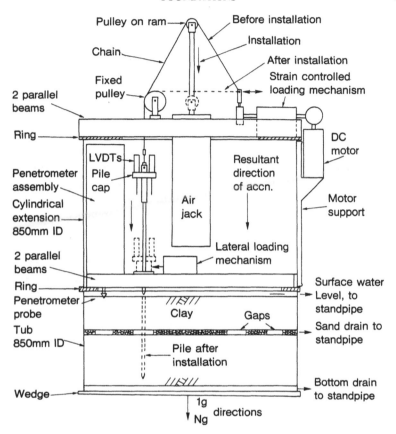

Figure 6.48 Push-in pile installation system. (After Nunez and Randolph, 1984. Balkema, with permission.)

(Nunez and Randolph, 1984). This system made use of the high self-weight of the mechanical components to effect pile installation. A pile was supported by a chain that passed over two pulleys and was then attached to the end of the loading system. One pulley was fixed and the other was attached to the end of the ram of an air jack which acted in an upward direction. During the consolidation period in flight, the air jack was pressurized and the ram fully extended, thereby supporting the pile above the soil. To install the pile, the air jack pressure was released rapidly and the ram retracted fully. The pile moved into the soil under its own weight and that of the pile cap. Later this system was modified to accommodate a double-acting pneumatic hammer to install the pile by impact force (Nunez *et al.*, 1988a). This was accomplished by means of five-way solenoid valves, operated at up to 2.5 Hz.

Further development of driving installation was made at the University of Colorado in which a hammer had a rotary valve, which allowed a pile to be driven at 25 *g* with an impact frequency of 27.7 Hz (Cyran *et al.*, 1991). At the corresponding prototype scale, the pile was being driven at a driving frequency of 67 blows per minute which is comparable to prototype pile installation.

6.3.3.3 Rate of installation.   Both rates of installation and of loading should be considered in modelling, particularly for cohesive soils. The rate of installation of a model pile in cohesive soils may be evaluated by the scaling for the dissipation of excess pore pressure, provided that the comparable value of coefficient of one-dimensional consolidation in a horizontal direction is used. In prototype situations, a time period is allowed after installation for excess pore pressures to dissipate before any major loading. Thus the measurement of excess pore-water dissipation in flight is necessary to ensure the establishment of equilibrium before loading is commenced.

6.3.3.4 Instrumentation.   Some pile tests may require measurements of shaft strains as well as pile tip resistance and deflection, from which the distribution of axial shaft resistance and bending moment along the pile can be determined. The information may be obtained by dividing the pile into several sections with load cells between each section (Cook and Lewis, 1980). More common arrangements use a load cell near the pile head and a series of strain gauges mounted on the surface of the piles. Generally 10–12 pairs of strain gauges will be needed along the pile depth in order to obtain sufficient information about the distribution of bending moment. Strain gauges may be attached to either the internal or external surface of the pile. Internal instrumentation implies that the pile will have a closed end in order to protect the instrumentation. External instrumentation allows the pile to be open-ended but requires a protective coating over the instrumentation. Details of gauge instrumentation are discussed by Nunez *et al.* (1988a).

6.4   Other foundations: pile foundation systems

6.4.1   Piled bridge abutment

Construction of an embankment behind piled bridge abutments on soft clay deposits leads to lateral deformation of the soil. Difficulties of dealing with the behaviour of piles in soil undergoing such movement are twofold: (i) how to estimate the ground movement and (ii) how to analyse the soil–pile interaction. Such difficulties may be overcome by centrifuge modelling (Springman *et al.*, 1991; Stewart *et al.*, 1991), from which useful experimentally deduced relationships or a new design method might be obtained.

Springman *et al.* (1991) used a two-layer system of soft clay overlying a sand or stiffer clay layer, in which a single row of free-headed piles or a pile group was installed. They examined the effect of pile top fixity and provided the details of bending moment distributions of each pile in a pile group. In their tests, a uniform surcharge pressure was applied by using a rubber bag. More realistic modelling of construction of an embankment in flight was made using a sand hopper (Stewart *et al.*, 1991). Two interesting observations were presented. Firstly, the maximum bending moment increased substantially after the embankment surcharge load exceeded about $3c_u$ and, secondly, the logarithm of the maximum bending moment increased linearly with embankment loading intensity. These findings compared favourably with field data. Similar trends have been observed for maximum lateral displacement under the embankment.

6.4.2 Pile–raft foundation

The pile–raft foundation, a combination of a raft foundation with piles, has been demonstrated as being particularly suitable for high-rise buildings. The behaviour of a vertically loaded pile–raft foundation on saturated overconsolidated clay has been investigated by centrifuge modelling (Thaher and Jessberger, 1991). The 58.8 m square pile–raft foundation with 64 piles of the 256 m high Fair Tower in Frankfurt was modelled.

The model raft was made of aluminium alloy and had a rough base; the tests were conducted at 150 g. At prototype scale the raft was 27 m wide with piles of 2.5 m diameter and 20–47 m in length. An overconsolidated clay body was made by pre-consolidation in the laboratory with subsequent reconsolidation in centrifuge flight. The pile–raft foundation model was installed after the pre-consolidation. After approximately 80% reconsolidation, a vertical loading test was carried out. Measurements included contact pressures at the underside of the raft, pile base resistance and strains along the pile.

It is of interest to see how the total load distributed over the piles changed

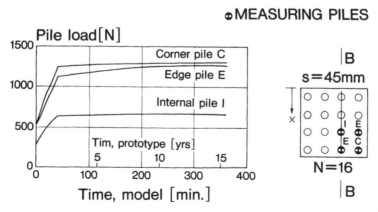

Figure 6.49 Changes in pile load with time for various locations. (After Thaher and Jessberger, 1991. Balkema, with permission.)

with time. Figure 6.49 shows an example of the data for the case of 16 piles in the base. As expected by the theory of elasticity, the corner and edge piles carried the greater load; the trend continued with time.

The ratio of the total pile load (Q_p) to the total applied load (Q) provides useful information for the design of a pile–raft foundation. Figure 6.50 shows the change of the ratio of Q_p/Q with various factors such as the number of piles, pile length, pile diameter and pile spacing. It can be seen from the figure that the total pile load depends on pile diameter and pile spacing and the influence of the number of piles is more significant than the pile length.

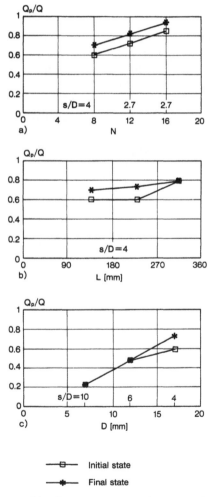

Figure 6.50 Apportionments of bearing pressure of pile for various factors. (After Thaher and Jessberger, 1991. Balkema, with permission.)

6.5 Summary

This chapter has described centrifuge modelling of various problems of shallow and pile foundations in both cohesionless and cohesive soils. Successful attempts at modelling of models were presented and direct comparisons between full-scale loading tests and corresponding centrifuge tests were given. Some useful information about modelling techniques was summarized and areas for further research were identified.

References

Aiban, S.A. and Znidarcic, D. (1991) Shallow footings on sand under vertical central, eccentric and inclined loads. *Centrifuge '91*, pp. 201–208. Balkema, Rotterdam.

Bagge, G., Fuglsang, L., James, R.G., Tan, T., Corté, J.F., Fargeix, D. and Garnier, J. (1989) Surface footings on a sand with a capillary zone. *Proc. 12th Int. Conf. Soil Mech.* Vol. 2, pp. 887–890. Balkema, Rotterdam.

Barton, Y.O. (1984) Response of pile groups to lateral loading in the centrifuge. *Proc. Application of Centrifuge Modelling to Geotechnical Design*, pp. 456–472. Balkema, Rotterdam.

Bolton, M.D. and Lau, C.K. (1988) Scale effect arising from particle size. *Centrifuge '88*, pp. 127–134. Balkema, Rotterdam.

Bouafia, A. and Garnier, J. (1991) Experimental study of p–y curves for piles in sand. *Centrifuge '91*, pp. 261–268. Balkema, Rotterdam.

Broms, B. (1964) Lateral resistance of piles in cohesionless soils. *J. Soil Mech. Found., ASCE*, SM3, 123–149.

Cook, D. and Lewis, R.T. (1980) Discussion: the use of physical models in design. *Proc. 7th Eur. Conf. Soil Mech. Found. Eng.*, **4**, 349–353.

Corté, J.F., Fargeix, D., Garnier, J., Bagge, G., Fuglsand, L., James, R.G., Shi, Q. and Tan, F. (1988) Centrifugal modelling of the behaviour of a shallow foundation—A cooperative test programme. *Centrifuge '88*, pp. 325–336. Balkema, Rotterdam.

Craig, W.H. (1984) Installation studies for model piles. *Proc. Application of Centrifuge Modelling to Geotechnical Design*, pp. 440–455. Balkema, Rotterdam.

Cyran, T.C., Mehle, J.S. and Goble, G.G. (1991) Centrifuge modelling of piles, *Centrifuge '91*, pp. 377–384. Balkema, Rotterdam.

Davis, E.H. and Booker, J.R. (1973) The effect of increasing strength with depth on the bearing capacity of clays. *Géotechnique*, **23**(4), 551–563.

Dean, E.T.R., James, R.G., Schofield, A.N., Tan, F.S.C. and Tsukamoto, Y. (1993) The bearing capacity of conical footings on sand in relation to the behaviour of spudcan footings of jackups. *Predictive Soil Mechanics*, pp. 230–253. Thomas Telford, London.

de Beer, E.E. (1963) The scale effect in transposition of the results of deep sounding tests on the ultimate bearing capacity of piles and caisson foundations. *Géotechnique*, **13**(1), 39–75.

de Beer, E.E. (1967) Bearing capacity and settlement of shallow foundations on sand. *Proc. Symp. Bearing Capacity and Settlement of Foundations* (ed. A.S. Vesić), pp. 15–33. Duke University, Durham, North Carolina, USA.

Dickin, E.A. and Wei, M.J. (1991) Moment carrying capacity of short piles in sand. *Centrifuge '91*, pp. 277–284. Balkema, Rotterdam.

Finn, W.D.L., Barton, T.O. and Towhata, I. (1984) Dynamic lateral response of pile foundations: centrifuge data and analysis. *Proc. Application of Centrifuge Modelling to Geotechnical Design*, pp. 143–153. Balkema, Rotterdam.

Fujii, N., Kusakabe, O., Keto, H. and Maeda, Y. (1988) Bearing capacity of a footing with an uneven base on slope: Direct comparison of prototype and centrifuge model behaviour. *Centrifuge '88*, pp. 301–306. Balkema, Rotterdam.

Garnier, J. and Canepa, Y. (1991) Effects of different footing conditions on the ultimate bearing pressure. *Centrifuge '91*, pp. 209–216. Balkema, Rotterdam.

Gemperline, M.C. (1988) Coupled effects of common variables on the behaviour of shallow foundations in cohesionless soils. *Centrifuge '88*, pp. 285–292. Balkema, Rotterdam.

Gemperline, M.C. and Ko, H.Y. (1984) Centrifuge model tests for ultimate bearing capacity of footings on steep slopes in cohesionless soils. *Proc. Application of Centrifuge Modelling to Geotechnical Design* (ed. W.H. Craig), pp. 206–225. Balkema, Rotterdam.

Hamilton, J.M., Phillips, R., Dunnavant, T.W. and Murff, J.D. (1991) Centrifuge study of laterally loaded pile behaviour in clay. *Centrifuge '91*, pp. 285–294. Balkema, Rotterdam.

Hettler, A. and Gudehus, G. (1988) Influence of the foundation width on the bearing capacity factor. *Soils Found.*, 28 (4), 81–92.

Hoadley, P.J., Barton, Y.O. and Parry, R.H.G. (1981) Cyclic lateral load on model pile in a centrifuge. *Proc. 11th Int. Conf. Soil Mech. Found. Eng.*, Vol. 1, pp. 621–625.

James, R.G. and Shi, Q. (1988) Centrifuge modelling of the behaviour of surface footings under combined loading. *Centrifuge '88*, pp. 307–312. Balkema, Rotterdam.

Kimura, T. and Saitoh, K. (1981) On the comparison of centrifuge tests with large scale model tests. *Proc. 10th Int. Conf. Soil Mech. Found. Eng.*, Vol. 4, pp. 674–675. Balkema, Rotterdam.

Kimura, T., Kusakabe, O. and Saitoh, K. (1985) Geotechnical model tests of bearing capacity problems in a centrifuge *Géotechnique*, 35 (1), 33–45.

Kitazume, M. (1984) Influence of loading condition on bearing capacity and deformation. *Proc. Int. Symp. on Geotechnical Centrifuge Model Testing*, Tokyo. 149–151. Japanese Society for Soil Mechanics and Foundation Engineering.

Ko, H.Y., Atkinson, R.H., Goble, G.G. and Ealy, C.D. (1984) Centrifugal modelling of pile foundations. *Analysis and Design of Pile Foundations* (ed. J.R. Meyer), pp. 21–40. ASCE.

Kubo, K. (1965) Experimental study of the behaviour of laterally loaded piles. *Proc. 6th Int. Conf. Soil Mech. Found. Eng.*, Vol. 2, pp. 275–279.

Kusakabe, O. (1980) Centrifuge Model Tests of an Oil Tank. MPhil Thesis, Cambridge University.

Kusakabe, O. (1993) Application of centrifuge modelling to foundation engineering. *Ground and Foundations*, Chugoku branch of Japanese Society of Soil Mechanics and Foundation Engineering, Vol. 11, pp. 1–10. (In Japanese.)

Kusakabe, O., Kimura, T. and Yamaguchi, H. (1981) Bearing capacity of slopes under strip load on the top surfaces. *Soils Found.*, 21 (4), 29–40.

Kusakabe, O., Yamaguchi, H. and Morikage, A. (1991) Experiment and analysis on the scale effect of N_γ for circular and rectangular footings. *Centrifuge '91*, pp. 179–186. Balkema, Rotterdam.

Kusakabe, O., Maeda, Y. and Ohuchi, M. (1992a) Large-scale loading tests of shallow footings in pneumatic caisson. *J. Geotech. Eng.*, *ASCE*, 118 (11), 1681–1695.

Kusakabe, O., Hagiwara, T., Maeda, Y. and Ohuchi, M. (1992b) Centrifuge modelling of in situ loading tests using undisturbed scoria samples. *Geotechnical Engineering*, *Proc. Japanese Society of Civil Engineers*, No. 457, pp. 107–116. (In Japanese.)

Kusakabe, O., Maeda, Y., Ohuchi, M. and Hagiwara, T. (1993) Attempts at centrifugal and numerical simulations of a large-scale in situ loading test on a granular material, *Predictive Soil Mechanics*, pp. 404–420. Thomas Telford, London.

Kutter, B.L., Abghari, A. and Shinde, S.B. (1988a) Modelling of circular foundations on relatively thin clay layers. *Centrifuge '88*, pp. 337–344. Balkema, Rotterdam.

Kutter, B.L., Abghari, A. and Cheney, J.A. (1988b) Strength parameters for bearing capacity of sand. *J. Geotech. Eng.*, *ASCE*, 114 (4), 491–498.

Lyndon, A. and Pearson, R.A. (1988) Skin friction effects on laterally loaded large diameter piles in sand. *Centrifuge '88*, pp. 363–370. Balkema, Rotterdam.

Matlock, H. (1970) Correlations for design of laterally loaded piles in soft clay. *Proc. 2nd Offshore Technology Conference*, OTC1204. Vol. 1, pp. 577–594.

Mikasa, M., Takada, N. and Yamada, K. (1973) Significance of centrifuge model test in soil mechanics. *Proc. 8th Int. Conf. Soil Mech. Found. Eng., Moscow*, Vol. 1.2, 273–278.

Nakase, A. (1966) Bearing capacity in cohesive soil stratum. *Report of PHRI*, 5 (12), 24–42. (In Japanese.)

Nakase, A., Kusakabe, O. and Wong, S.F. (1984). Centrifuge model tests on bearing capacity of clay. *J. Geotech. Eng.*, 110 (12), 1749–1765. ASCE.

Nakase, A., Kimura, T., Saitoh, K., Takemura, J. and Hagiwara, T. (1987). Behaviour of soft clay with a surface crust. *Proc. 8th Asian Regional Conf. Soil Mech. Found. Eng.*, Vol. 1, pp. 401–404.

Nunez, I.L. and Randolph, M.F. (1984) Tension pile behaviour in clay—centrifuge modelling techniques. *Proc. Application of Centrifuge Modelling to Geotechnical Design*, pp. 87–102. Balkema, Rotterdam.

Nunez, I.L., Hoadley, P.J., Randolph, M.F. and Hulett, J.M. (1988a) Driving and tension loading piles in sand on a centrifuge. *Centrifuge '88*, pp. 353–362. Balkema, Rotterdam.

Nunez, I.L., Phillips, R. Randolph, M.F. and Wesselink, B.D. (1988b) Modelling laterally loaded piles in calcareous sand. *Centrifuge '88*, pp. 371–384. Balkema, Rotterdam.

Oda, M. and Koishikawa, I. (1979) Effect of strength anisotropy on bearing capacity of shallow footings in a dense sand. *Soils Found.*, **19** (3), 15–28.

Oldham, D.C.E. (1984) Experiments with lateral loading on single piles in sand. *Proc. Application of Centrifuge Modelling to Geotechnical Design*, pp. 122–142. Balkema, Rotterdam.

Ovesen, N.K. (1975) Centrifuge testing applied to bearing capacity problems of footings on sand. *Géotechnique*, **25**, 394–401.

Ovesen, N.K. (1980) Discussion: The use of physical models in design. *Proc. 7th Eur. Regional Conf. Soil Mech. Found. Eng.*, Brighton, Vol. 4, pp. 315–323. British Geotechnical Society, London.

Poulos, H.G. and Davis, E.H. (1980) *Pile Foundation Analysis and Design*. Wiley, New York.

Pu, J.L. and Ko, H.Y. (1988) Experimental determination of bearing capacity in sand by centrifuge footing tests. *Centrifuge '88*, pp. 293–300. Balkema, Rotterdam.

Sabagh, S.K. (1984) Cyclic axial load tests on piles in sand. *Proc. Application of Centrifuge Modelling to Geotechnical Design*, Manchester, pp. 103–121. Balkema, Rotterdam.

Scott, R.F. (1981) Pile testing in a centrifuge. *Proc. 10th Int. Conf. Soil Mech. Found. Eng.*, Vol. 2, pp. 839–842.

Shields, D.H., Scott, J.D., Baner, G.E., Deschenes, J.H. and Barsvary, A.K. (1977) Bearing capacity of foundations near slopes. *Proc. 9th Int. Conf. Soil Mech. Found. Eng.*, Vol. 1, pp. 715–720.

Siddiquee, M.S.A., Tanaka, T. and Tatsuoka, F. (1992) A numerical simulation of bearing capacity of footing on sand. *Proc. 27th Annu. Meet. Japanese Soc. Soil Mech. and Found. Eng.*, Vol. 2, pp. 1413–1416.

Springman, S.M., Randolph, M.F. and Bolton, M.D. (1991) Modelling the behaviour of piles subjected to surcharge loading. *Centrifuge '91*, pp. 253–260. Balkema, Rotterdam.

Stewart, D.P., Jewell, R.J. and Randolph, M.F. (1991) Embankment loading of piled bridge abutments on soft clay. *Geocoast '91*, Vol. 1, pp. 741–746.

Tatsuoka, F., Okahara, M., Tanaka, T., Tani, K., Morimoto, T. and Siddiquee, M.S.A. (1991) Progressive failure and particle size effect in bearing capacity of a footing on sand. *Geotech. Eng. Congr., ASCE*, Vol. II, pp. 788–802.

Terashi, M., Kitazume, M. and Kawabata, K. (1989) Centrifuge modelling of a laterally loaded pile. *Proc. 12th Int. Conf. Soil Mech. Found. Eng.*, Vol. 2, pp. 991–994.

Terashi, M., Kitazume, M. and Tanaka, H. (1984) Application of PHRI geotechnical centrifuge. *Proc. Int. Symp. on Geotechnical Centrifuge Model Testing*, Tokyo, pp. 164–171. Japanese Society for Soil mechanics and Foundation Engineering.

Terashi, M., Kitazume, M., Maruyama, A. and Yamamoto, Y. (1991) Lateral resistance of a long pile in or near the slope. *Centrifuge '91*, pp. 245–252. Balkema, Rotterdam.

Thaher, M. and Jessberger, H.L. (1991) The behaviour of pile–raft foundations investigated in centrifuge model tests. *Centrifuge '91*, pp. 225–234. Balkema, Rotterdam.

Tomlinson, M.J. (1986) *Foundation Design and Construction*, 5th edn. Longman, London.

Wesselink, B.D., Murff, J.D., Randolph, M.F., Nunez, I.L. and Hyden, A.M. (1988) Analysis of centrifuge model test data from laterally loaded piles in calcareous sand, *Proc. Int. Conf. Calcareous Sediments, Perth*, Vol. 1, pp. 261–270. Balkema, Rotterdam.

Yamaguchi, H., Kimura, T. and Fuji-i, N. (1976) On the influence of progressive failure on the bearing capacity of shallow foundations in dense sand. *Soils Found.*, **16** (4), 11–22.

Yamaguchi, H., Kimura, T. and Fujii, N. (1977) On the scale effect of footings in dense sand. *Proc. 9th Int. Conf. Soil Mech.*, Vol. 1, pp. 795–798.

Vesic, A.S. (1963) Bearing capacity of deep foundation in sand. *Highway Res. Rec.*, **39**, 112–153.

7 Dynamics

R.S. STEEDMAN and X. ZENG

7.1 Introduction

Earthquake and blast are the two areas which have dominated the use of the centrifuge for modelling dynamic problems. Although other specialist areas involving dynamics have also been addressed, such as vibrations of structures, pile driving and wave-induced cyclic loading, strong international academic interest in earthquake induced liquefaction, on the one hand, and military interest in studying weapons effects, on the other, has dictated the pace of development in centrifuge capability for dynamic modelling.

The application of the centrifuge to problems involving weapons effects was recognised early on by the Soviets but only adopted much later in the west. By the 1940s the centrifuge was being used routinely in the USSR to address problems involving cratering and blast loading, where the cube power scaling for energy (Table 7.1) provides a potent modelling tool. It is thought that over the ensuing decades the centrifuge provided a critical tool in the Soviet understanding of weapons effects and consequently had a significant influence on strategy. In the 1970s similar experiments were conducted in the UK and in the USA as researchers turned to the centrifuge as an alternative to expensive and problematic field testing. Their findings showed clearly the advantage of centrifuge testing over field testing for studying weapons effects.

Since around 1980 there has been a rapid development of interest world-wide in centrifuge modelling of earthquake problems. A number of centrifuge centres in different countries around the world have acquired the capability of

Table 7.1 Key scaling relationships for dynamic modelling

Parameter	Field	Centrifuge model (Ng)
Stress	σ	σ
Displacement	x	x/N
Velocity	v	v
Acceleration	A	NA
Time (dynamic)	t	t/N
Frequency	f	Nf
Time (consolidation)	T_v	T_v/N^2
Energy	E	E/N^3

earthquake centrifuge modelling. A wide range of research projects has been undertaken. Data from centrifuge tests have helped engineers to understand the mechanisms underlying geotechnical earthquake engineering problems, to verify numerical codes and to assist with the design of structures in the field. The principles of dynamic centrifuge modelling are now well understood. The interpretation of model tests has changed gradually from seeking mainly qualitative explanations towards using quantitative analyses using detailed information from miniature transducers capable of responding in the frequency ranges of interest. Recent developments in earthquake centrifuge modelling have included the design of robust earthquake actuators, the design of model containers to minimise the influence of undesirable boundary effects and the combination of earthquake centrifuge modelling with other modelling approaches.

7.2 Dynamic scaling and distortions in dynamic modelling

In geotechnical centrifuge modelling it is common to adopt a length scale that is inversely proportional to the gravity scale: thus a $1/N$ scale model is tested at Ng. (In other fields, such as fluid mechanics, this may not always be the case.) Accepting this approach the important scaling relationships for dynamic centrifuge modelling are presented in Table 7.1.

A new time scale emerges from this approach in considering time for inertial events. A dynamic perturbation:

$$x = a \sin \omega t \tag{7.1}$$

in the field is properly represented by

$$x_{\mathrm{m}} = (a/N) \sin \sqrt{N} \, \omega t \tag{7.2}$$

in a $1/N$ linear scale model experiment conducted at $1\,g$, but by

$$x_{\mathrm{Nm}} = (a/N) \sin N \, \omega t \tag{7.3}$$

in a $1/N$ scale model experiment at Ng. This arises because the accelerations applied in a $1\,g$ model are unchanged from those in the field event, whereas accelerations in a model at Ng are increased by the factor N. Thus in a centrifuge model the period of the dynamic perturbation is reduced by N bringing immediate conflict with the alternative standard interpretation of time in the centrifuge model based on diffusion, where time in the model is apparently reduced by N^2.

Thus if the same soil and pore fluid as in the field are used in the model, the time for consolidation would be scaled down by N^2 while the duration of a dynamic event is scaled down by N. In many geotechnical problems the ability to separate the response into drained and undrained behaviour, which are considered to be essentially uncoupled, avoids this conflict of time scaling.

However, where the undrained and drained response is coupled, for example in the earthquake generation of pore pressure in fine sands, then careful consideration must be given to harmonising these two scales.

There are two routes which are commonly adopted to achieve a unique time scale, and these are either to increase the viscosity of the pore fluid or to reduce the particle size of the soil so as to reduce its permeability.

Increasing the viscosity of the pore fluid by N is readily achieved using either silicone oil or a glycerine/water mix. Whilst an increase in viscosity of 50 or 80 times over that of water may sound high, in practice such a fluid still appears to be much more like water than like, say, engine oil. Models are generally built dry and saturated with the pore fluid under vacuum. Difficulties can arise if the model contains impermeable regions or layers, or a mixture of clays and sands, and it may be necessary to tilt the model or to saturate in stages to avoid entrapment of air. Clays, either from block samples, recompacted natural clays or remoulded modelling clays can only practically be saturated with water.

The second alternative, that of reducing grain size, can be effective if it is necessary for other reasons to use water as the pore fluid. Water saturated models are generally easier and quicker to build and to handle. For many fine sands the permeability can be considered as a function of the square of the D_{10} size, the grain size defining the finest 10% by mass from a sieve analysis. Thus to reduce the permeability by N would require reducing the grain size by $\sqrt{N}$, which in many cases may be possible, particularly for low g testing. However, it is important to recognise that altering grain size will also alter the mechanical properties of the soil, particularly where high stresses are involved and particle crushing on asperities may be significant.

Energy scales with N^3. This arises by analogy with the scale for mass which is also N^3; it provides a powerful opportunity for studying the effects of large explosions without the environmental consequences and practical difficulties of field testing. Thus 1 g of high explosive at $100\,g$ will generate a crater identical to 1 tonne of high explosive in the field, but in miniature. This relationship provides the basis for all modelling of non-nuclear or simulated nuclear events.

The high velocities associated with explosions, velocities which may be of the same order as the velocity of the centrifuge platform in flight, bring distortions to the model from Coriolis effects. Although Coriolis is often raised as a major hurdle, this is largely due to misunderstanding as both the phenomenon and its effects are easily predicted and therefore can be readily accommodated in any interpretation of a model test.

The problem for the observer is best illustrated by example. Suppose the observer drops a stone down a deep shaft towards the centre of the Earth. Neglecting air resistance, does the particle fall straight down the hole? No, but not because of any force accelerating the particle towards the wall tangentially, but simply because the shaft is moving. The particle leaves the observer's hand with a tangential velocity associated with the rotation of the Earth at the Earth's surface. The particle is accelerated towards the centre of the Earth by gravitational attraction and acquires a radial velocity. But its tangential velocity is constant and exceeds the tangential velocity of the hole at any depth. Thus the observer sees the stone move 'forward' and it hits the leading side of the shaft. Seen from the stone's perspective, it was quietly falling through the universe towards the centre of the Earth when the side of the shaft jumped up and hit it, as the driver of the crashed car said to the police officer.

In fact this example is closely analogous to the situation in the centrifuge, with the exception that particles in free space are not accelerated towards the central axis of the machine. Coriolis 'acceleration', like centrifugal acceleration, is entirely fictitious; such fictitious accelerations are not a result of a physical interaction but are invoked by observers in one, Newtonian, reference system to explain the apparently curious movements of particles in a second system accelerating relative to their own. Particles ejected or released from a centrifuge model in flight travel with a constant horizontal velocity in a straight line in the direction in which they were set in motion (neglecting air resistance) until they strike the centrifuge again or the wall of the centrifuge chamber. They feel no Coriolis acceleration or centrifuge acceleration.

Expressed vectorially, the equation of motion for a particle of mass m in our inertial system is simply:

$$F = ma_s$$

where a_s is the acceleration of the particle relative to our 'fixed' space system. However, to interpret the motion of a particle relative to a system rotating at constant angular velocity ω, the equation of motion becomes:

$$F - 2m(\omega \times v_r) - m\omega(\omega \times r) = ma_r$$

where r, v_r and a_r are the position, velocity and acceleration vectors of the particle relative to the rotating system. The two additional terms in the equation of motion represent the Coriolis force and the centrifugal force respectively. The Coriolis force is proportional to the cross-product of ω and v_r and is of magnitude

$$|F_c| = 2m|\omega||v_r|\sin\theta$$

with its direction following a right-angle rule, perpendicular to both ω and v_r. Clearly if ω and v_r lie in the same direction or if the particle remains at rest in the rotating frame the Coriolis force is zero.

The maximum Coriolis acceleration $A_c = 2\omega V$ may be expressed, instantaneously, in terms of a radius R such that $A_c = V^2/R$, where $R = V/2\omega$. Comparing this acceleration with the magnitude of the constant centrifugal acceleration $a = v^2/r = v\omega$ gives the Coriolis 'error':

$$A_c/a = 2V/v$$

Clearly if V becomes greater than 5% v, for example, then the error in neglecting the Coriolis term exceeds 10%. However, if the velocity is very high, for example $V > 2v$, then the particle is travelling in a path (relative to the model) which is staighter than the curvature of the model itself and other errors are likely to be more significant.

The analysis of the trajectory of a particle, as seen from the model chamber, shows why ejecta from a crater are thrown predominantly forwards in flight. Here the Coriolis effect is most easily seen by considering the actual path of the particle (which is a straight line in the horizontal plane) and transforming that path into coordinates local to the model itself.

Consider a particle which is ejected at a velocity V relative to the package and at an angle ϕ from the instantaneous direction of travel of the package. The actual path taken by the particle will be a straight line which is the vector sum of the tangential velocity of the package and the velocity of the particle relative to the package, V. Figure 7.1 shows the trajectory of a particle and the local x,y coordinate system defined relative to the package. Are there particles which land back at the origin, $x = 0$, $y = 0$? For this condition the time taken for a particle to reach the perimeter again will equal the time taken for the origin to move around the perimeter:

$$t = \frac{2R \sin (\theta/2)}{V_{\text{abs}}} = \frac{R\theta}{\omega R}$$

giving the condition to land at 0,0:

$$V_{\text{abs}} = \frac{2\omega R}{\theta} \sin \left(\frac{\theta}{2}\right)$$

where V_{abs} is the absolute velocity of the particle, and $\theta/2$ is the angle the actual path the particle makes to the initial x-axis. Considering V_x and V_y as

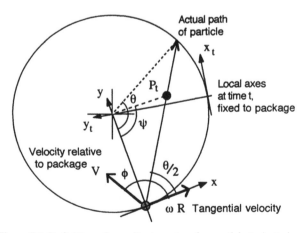

Figure 7.1 Definition of coordinate system for particle trajectories.

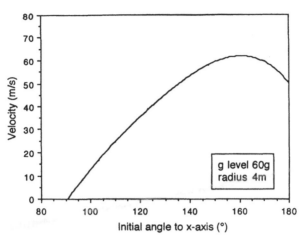

Figure 7.2 Ejector velocity as a function of angle of ejection to land at 0,0.

the x and y components of V_{abs}, then this condition can also be expressed as follows:

$$V_x = \frac{\omega R}{\theta}(\sin\theta - \theta)$$

$$V_y = \frac{\omega R}{\theta}(1 - \cos\theta)$$

Note that $|V_{abs}| = \sqrt{\{(\omega R + V_x)^2 + V_y^2\}}$ where v_x and v_y are components of the initial velocity V, such that $|V| = \sqrt{V_x^2 + V_y^2}$. Selecting a range of values of θ from 0–180° and equating components gives the necessary values of $|V|$ for the particle to land at 0,0. The initial ejector angle that V makes to the x-axis is given by $\phi = \cos^{-1}(V_x/|V|)$. Figure 7.2 shows the relationship between V and ϕ for a package subjected to a radial acceleration of $60\,g$ at a radius of $4\,\text{m}$. Notice that in order to land back at 0,0 the particle must be ejected 'backwards', i.e. at an angle greater than 90° relative to the package. This is necessary in order to counter the large tangential velocity and 'slows up' the particle in the initial x-direction. The y-component of the velocity then carries it across the circular path to meet the package again.

In local Cartesian coordinates the particle's position relative to the tub for a general trajectory may be traced by transforming the global coordinates of the particle into the local coordinates at the current location of the package on the perimeter at a time t in Figure 7.1, for example. The transformation of coordinates from x,y to x_t,y_t is affected by a rotation and translation of the coordinate system. The transformation is therefore given by:

$$x_t = x\cos\psi + y\sin\psi - x_T\cos\psi - y_T\sin\psi$$
$$y_t = -x\sin\psi + y\cos\psi + x_T\sin\psi - y_T\cos\psi$$

where $\psi = \omega t$ is the rotation of the axes and x_T, y_T are the translation components. From Figure 7.1 it is clear that:

$$x_T = R \sin \psi$$

$$y_T = R(1 - \cos \psi)$$

The coordinates of the particle x, y during its flight relative to the global system are given by:

$$x = V_{abs} t \cos (\theta/2)$$

$$y = V_{abs} t \sin (\theta/2)$$

where V_{abs} and θ were defined above.

Figure 7.3 shows the theoretical trajectories of particles which land back at 0,0 for different combinations of V and ϕ based on an arbitrary selection of actual paths at angles of $\theta = 350°, 270°, 180°, 90°$ and for a range of ejector angles from 179.9° to 119.7°. Consider the case of $V = 57.6 \,\text{m/s}$, $\phi = 147.5°$. The x- and y-components of the velocity V are then $V_x = -48.6 \,\text{m/s}$ and $V_y = 30.95 \,\text{m/s}$. However the tangential velocity at $60 g$ at $4 \,\text{m}$ radius is simply $\omega R = 12.15 \times 4 = +48.6 \,\text{m/s}$, equal and opposite to the V_x component. The actual path followed by the particle is therefore along a diameter of the circle. We can deduce quite simply that its maximum 'height' above the surface of the model should therefore be equal to the radius $R = 4 \,\text{m}$. This is confirmed in Figure 7.3.

Figure 7.4 shows trajectories of particles with a fixed ejector velocity V but varying ejector angles ϕ. What is clear is the bias by which particles are thrown 'forwards' in the direction of travel because of their high absolute velocity but shorter travel paths than the package 'below'. This is readily seen in practice: Figure 7.5 shows sand ejected from a crater lying around the

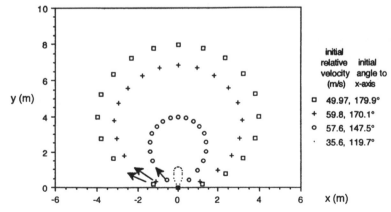

Figure 7.3 Particle trajectories which land at 0,0 for a range of ejector angles and velocities (g level $60 \,g$; radius $4 \,\text{m}$).

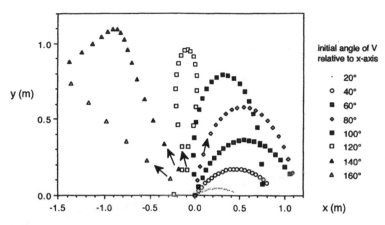

Figure 7.4 Particle trajectories with fixed ejector velocity, $V = 32.12\,\text{m/s}$.

Figure 7.5 Ejecta thrown forward in blast model experiment.

leading side of the model chamber. Distortion of the ejecta does not necessarily mean distortion of the crater, however, as the velocities of particles
which do not 'escape' from the surface are clearly much less than those that
do.

It is important, finally, to remember that despite the curious loops which
mark the apparent trajectories of some of the ejected particles, the true paths
are straight in the horizontal (x,y) plane. Particles which are not ejected, of
course, will continue to accelerate and a more complex analysis will be
necessary. In a rigorous analysis, the Earth's gravity would also need to be
considered, as would air resistance in flight. In addition, beyond the protected
domain of the centrifuge chamber, above and below the package there will be
very strong winds created by the rotation of the centrifuge itself. Thus
although the analysis presented here is theoretically correct, it serves best to
illustrate the principles of Coriolis and to provide some confidence that the
phenomenon is understandable.

A second major potential source of error in dynamic modelling commonly
arises in earthquake modelling, in which the elastic wave energy to excite the
model needs to be input through the floor and walls of the model chamber. In
the field, the soil horizon is continuous and just as in numerical modelling, a
physical model must endeavour to provide a lateral boundary which simulates as far as possible the free field condition. Any boundary must as far as
possible minimise distortions induced in the soil stress field, the soil strain
field, and from seismic waves (Schofield and Zeng, 1992).

A soil column subjected to a base shear distorts in shear rather than in
bending because on each side of the column complementary shear stresses can
develop which permit the rotation of principal stresses required for the
propagation of a shear wave. To simulate the free field condition on the
boundary, therefore, requires not only the matching of the lateral displacement with depth in the soil column but also the development of complementary shear stress on vertical planes to simulate the infinite stress field
condition. To minimise reflected seismic wave energy requires matching of the
impedance of the boundary with the soil. Many different approaches have
been adopted to provide an 'ideal' boundary. This is discussed in more detail
below.

With this background in mind we shall now turn to specific examples of the
use of the centrifuge for dynamic modelling applications.

7.3 Earthquake modelling

7.3.1   *The role of the centrifuge in earthquake engineering*

The need for earthquake centrifuge modelling is closely associated with the
nature of earthquakes in the field. Very little quantitative field data exist of the

response of soil deposits or soil structures to strong ground shaking, although it is well established that soil behaviour can play a key role in dynamic structural response.

Centrifuge model tests using a shaking table in flight, however, provide an opportunity to observe the overall soil–fluid–structure response with a versatility that is completely unattainable in the field or in cyclic or dynamic laboratory element testing.

The opportunity to study the phenomenon of earthquake-induced liquefaction has been the principal driving force behind the widespread interest in generating strong ground shaking on a centrifuge model. More recently, earthquake modelling has focused on the behaviour of soil–structure systems, particularly in the presence of high excess pore pressures leading to a degradation of stiffness and strength in the soil. This modelling has allowed an exploration of problems too complex to visualise on paper or by computer.

Early work in Japan used a tilting mechanism to induce an inclined acceleration field on models (Mikasa et al., 1969), but it was not until the late 1970s and into the 1980s that shaking systems on centrifuges became truly dynamic in their capabilities. Since then considerable efforts have been made by researchers throughout the world to enhance the role and to extend the capability of centrifuge modelling in geotechnical earthquake engineering (Schofield, 1981; Scott, 1983; Schofield and Steedman, 1988). More than ten of the centrifuge centres in Europe, the USA and Japan have the capability of earthquake centrifuge modelling with new facilities being developed on several other machines.

By 1985 a comprehensive study on liquefaction related problems, sponsored by the US National Research Council's Committee on Earthquake Engineering (1985) had concluded that

centrifuge tests can provide valuable insights in understanding the behaviour of foundations and earth structures and in formulating and testing mathematical models to predict performance.

As for the role of centrifuge modelling in design, the report recommended that:

it appears at present that there is no direct role for centrifuge model testing in conventional engineering projects. However, if a project warrants a considerable theoretical analysis, and especially if theoretical methods still in the development stage are to be used, then centrifuge model testing to validate the theoretical analysis and perhaps to provide concepts for design is worthy of consideration.

Since that date both the experience base and the quality of data available from earthquake centrifuge modelling have improved significantly. The role of centrifuge modelling has been extended and enhanced and case studies have demonstrated that dynamic centrifuge modelling has an important role to play in geotechnical earthquake engineering design.

One potential alternative to centrifuge tests is the shaking table, which is widely used for dynamic structural research. Shaking table tests at $1\,g$ can often provide a more complex input than is possible on a centrifuge, but they are incapable of providing the depth of equivalent prototype which is often necessary to investigate fully the behaviour of a dynamic soil–fluid–structure system. At high stress ratio also, non-linearities in the soil behaviour will not be capable of interpretation at scales much greater than 1. Nevertheless, the shaking table test may be the preferred option at low scales, for example between 1 and around 10, depending on the type of problem under consideration. Iai (1989) discussed in detail possible scaling relations that can be adopted for such studies. Beyond this the distortions introduced by $1\,g$ modelling outweigh the benefits of easy handling and large model volume and the centrifuge should be seen as the preferred option (Steedman, 1990).

The application of centrifuge modelling to earthquake problems is now fully developed and a number of important milestones have been achieved.

First, realistic data of the seismic response of a wide variety of earth structures and foundations have been generated. Comparisons with field experience of earthquake induced failures indicate that the mechanisms of behaviour observed in the models correspond closely to those of similar structures observed in the field. Failure of retaining walls under earthquake-induced lateral accelerations, for example, is widely seen in the field. Centrifuge studies of the failure of walls retaining cohesionless fill, for example in Figure 7.6, have shown clearly the steeper sliding surface in the backfill, the outward sliding displacement once a critical (threshold) acceleration is exceeded and the residual pressures noted in field cases. In addition, centrifuge models have shown the significance of the soil stiffness in determining amplification of motion and interaction between wall and soil, the nature of hydrodynamic pressures and the consequences of degradation of stiffness of the backfill on the response of the structure. These insights have led to a better understanding of the limitations of conventional analyses (such as the Mononobe–Okabe analysis) and allowed the development of new approaches.

Second, data from earthquake centrifuge models have been used for certain specific problems to enhance and in some cases, to validate, numerical codes, particularly several of the research codes which have been developed for dynamic effective stress analysis such as SWANDYNE (University College of Swansea), DYNAFLOW (Princeton University) and TARA (University of British Columbia). The application of centrifuge modelling to code validation for earthquake problems is the central theme of a major NSF funded research project, currently nearing completion, entitled VELACS (VErification of Liquefaction Analysis by Centrifuge Studies).

Third, earthquake centrifuge models are now an accepted part of the design process for large-scale projects where uncertainties exist over the application of conventional earthquake design approaches. As in other problems, the centrifuge model test is used to support the design approach by generating data of a similar class of problem to that facing the designer. However, this is

Figure 7.6 Failure of a monolithic wall retaining dry sand under earthquake shaking.

particularly important in the earthquake area because there is no opportunity to carry out field trials and there are very few documented case histories. The design brief for the Pier 400 scheme to develop the Port of Los Angeles, for example, required the designer to use centrifuge model testing in the valid-ation of the earthquake design. Similarly, the recent earthquake design of the dock walls for the nuclear submarine refit facility at Devonport in the UK used centrifuge model test data to support the design approach. Both of these schemes involve large capital investment and important economies could be achieved by reducing conservatisms in the earthquake design process.

7.3.2   *Techniques for achieving earthquake shaking on a centrifuge: earthquake actuators*

Generating earthquake-like shaking of a model in flight on a centrifuge requires a power source or actuator. This will have to operate under high gravity and to provide high peak power at high frequency. In practice the

towers, slopes, retaining walls, dykes, embankments and artificial islands. Although the input motion is limited to a 'tone burst' (shaking predominantly at a single frequency) with a fixed number of strong cycles the system has proved robust and very economical over its long life. The simple input motion, similar in character to the damaging cyclical motions often experienced on soft sites in the field, has proved invaluable to researchers attempting to unravel the modes and phase relationships in the complex dynamic response of soil–structure systems.

Generally speaking a mechanical shaker has the advantage of relatively low cost in manufacture and maintenance, being easy to operate and capable of generating sinusoidal vibration at a range of frequencies. The main limitation of mechanical systems is the narrow frequency band which is usually available.

Since the early 1980s considerable efforts have been made in the USA and Japan to develop servo-controlled electro-hydraulic systems. A hydraulic shaker obtains the energy for vibration from a hydraulic piston and has the capability to provide a nominally programmable input motion. The principles and basic design requirements of a hydraulic shaker were outlined by Ketcham *et al.* (1988). In the University of Colorado design a large hydraulic shaker was mounted under a shaking table with the swinging basket itself providing the reaction mass. The direction of motion induced by the actuator is parallel to the axis of the centrifuge itself, reducing the errors which may be introduced by Coriolis effects or, on small radius centrifuges, by the curvature of the centrifuge acceleration field.

The primary advantage of a hydraulic shaker is the adjustability in its input motion so that it can produce a range of different broader band input motions. This type of earthquake actuator has been adopted on a number of academic centrifuges in the USA, including Caltech, U.C. Davis, University of Colorado at Boulder, R.P.I., Princeton, and in Japan at the Port and Harbour Research Institute, Tokyo Institute of Technology and Kyoto University. So far hydraulic actuators have been used for only relatively small payloads and at g levels typically up to around $100\,g$. Large systems for high g operation are likely to be both costly and risky, given the present limitations in the operation of servo-hydraulic valves above around $100\,g$ or at high frequency.

7.3.3 Model containers to reduce boundary effects

The primary source of damaging earthquake energy in the field is elastic shear waves which propagate upwards from stiffer underlying strata, rotating the principal stress direction in each soil element by the imposition of shear stresses on planes perpendicular and parallel to the direction of wave propagation. For an idealised vertically propagating shear wave, the imposed stresses on a soil element in an infinite stratum are shown in Figure 7.7. In a centrifuge test a model of an earth structure is built inside a model container,

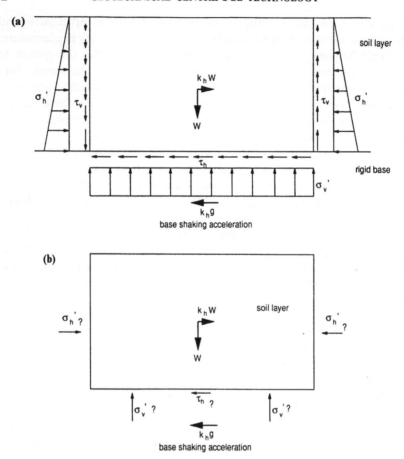

Figure 7.7 Distortion in stress field due to smooth end walls. (a) Distribution of stresses in a soil layer of infinite lateral extent; (b) distribution of stresses in a model with smooth rigid end walls.

through which the excitation has to be transmitted, thus imposing artificial boundaries and leading potentially to boundary effects.

Boundary effects have to be carefully addressed in order that a centrifuge test can realistically model the behaviour of the corresponding prototype (Schofield and Zeng, 1992). Three major boundary effects may be caused by a model containment, namely the effects on the stress field, on the strain field and on the seismic waves, as shown in Figures 7.7, 7.8 and 7.9. The design criteria for an ideal model container for earthquake centrifuge modelling should ensure the following:

1. The end walls function as shear beams with the same dynamic stiffness as the adjacent soil, so as to achieve strain similarity and to minimise the interaction between the soil and the end walls and hence minimise the generation of compression waves.

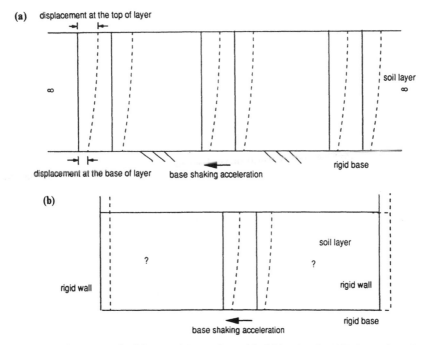

Figure 7.8 Deformation of soil in a model container with rigid end walls. (a) Deformation of soil shear beams in the prototype; (b) deformation of soil in the actual model.

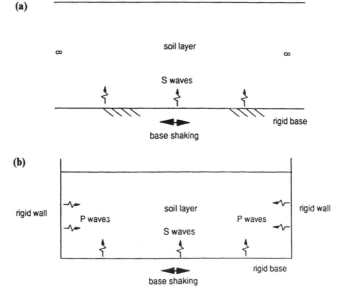

Figure 7.9 Earthquake waves in model and prototype. (a) S wave in the prototype; (b) S and P waves in the model.

2. Each end wall should have the same friction as the adjacent soil so that it can sustain the complementary shear stresses induced by base-shaking and thus the same stress distribution as in the prototype shear beam can be achieved.
3. The side walls should be frictionless so that no shear stress is induced between the side walls and soil during base-shaking to create the same two-dimensional condition as in the prototype.
4. The model containment should be rigid statically to achieve a zero lateral strain (K_o) condition, and after shaking to maintain its initial size.
5. The frictional end walls should have the same vertical settlement as the soil layer contained during the spin-up of a centrifuge to avoid initial shear stresses on the boundary.

Of these, the requirement for the end walls to function as shear beams (1) is the most difficult to satisfy since the stiffness of the soil may vary during the shaking, and this transient change may be large if liquefaction of a soil layer occurs. However, even if it is not possible to vary the stiffness of the end walls during a test, it should be possible to design end walls which approximately match the stiffness of soil over working conditions, in which the range of stress change is not large. During base-shaking, the same dynamic shear stiffness means that the end walls should have the same deflection and natural frequency as the soil in the model container.

The first model containers used had rigid smooth walls, similar to containers used for 'static' models. However, it was recognised that the rigid walls may cause stress and strain dissimilarities, and generate horizontally propagated P waves during base-shaking. Following an experimental and numerical study, Whitman (1988) showed that the effect of a rigid end wall extends to a distance of about twice the depth of the stratum.

To solve the problem of end effects, different types of boundaries have been developed for various kinds of problems. Energy absorbing materials such as Duxseal, first used at Princeton University, were proposed as an attempt to create an ideal end condition. The use of Duxseal clearly reduced the influence of incident waves. However, its compressibility and friction characteristics have never been satisfactorily determined in standard laboratory tests and this has posed problems in subsequent analyses of the data from such tests. Another solution, used at Cambridge University for comparison with numerical models, was free slopes which avoided the end wall altogether. These models clearly did not attempt to represent an infinite soil stratum but provided a well-defined boundary for subsequent dynamic finite element modelling.

The concept of a flexible boundary for earthquake centrifuge modelling was first reported by Whitman et al. (1981) who developed a stacked ring apparatus made from Teflon-coated aluminium rings held together using a system of tensioned springs mounted externally. Intended to simulate the behaviour of a column of soil within a stratum, the rings were expected to move with the soil and to develop dynamic shear stresses at the soil–ring interface, complementing the dynamic shear stresses on horizontal planes through the soil, and preventing extensional strain in horizontal directions. The friction between the rings was small so that the rings could move relative to each other, thus creating a flexible boundary.

However, the stacked-ring apparatus still had several unsatisfactory aspects. Without direct control of the lateral stiffness of the apparatus, the relative displacement between the rings could cause strain concentrations at certain levels of a soil column and potentially local arching in the soil. Secondly, the rings had a smooth inner surface which could not sustain the complementary shear stresses, leading to distortions in the stress field. Thirdly, the circular shape and the relatively low height-to-length ratio of early versions created a three-dimensional problem that was difficult to analyse. Finally, it was not clear that the container was acting as a shear column and not simply as a beam in bending.

The study of large deformations and liquefaction failures of slopes and retaining walls required a different shape of box and this led to a modification of the stacked ring concept, the laminar box, which was developed at Caltech (Hushmand *et al.*, 1988). The box consisted of rectangular frames stacked together with low-friction bearings between the frames, intended to mirror the stiffness of a liquefied soil column and to allow for large permanent deformation. Tests using the box showed that it satisfied the requirements for a flexible boundary but did not develop the necessary complementary shear stresses on the end walls.

Further development of the stacked ring concept by Schofield and Zeng (1992) led to the ESB (equivalent shear beam) container, which was built from rectangular frames of dural separated by rubber layers so as to achieve the same dynamic stiffness as the soil sample. To sustain the complementary

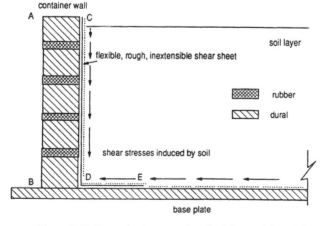

Figure 7.10 Shear sheet near end wall of the model container.

shear stresses induced by base-shaking, a flexible and inextensible friction sheet was attached to each end wall as shown in Figure 7.10. Data from model tests showed that the end walls had a deflection profile close to the predicted deformation of the soil column and that complementary shear stresses were sustained by the shear sheet, leading to uniform accelerations along the length of the model. This design appears to satisfy most of the requirements for a model container for earthquake centrifuge modelling although it would be necessary to tune the stiffness of the rubber gaskets carefully to ensure that an appropriate stiffness is achieved for the soil sample being tested. Different boxes may be needed for different g levels and clearly gross changes in soil stiffness, due to liquefaction for example, may be a source of error.

7.3.4 Modelling the effects of earthquakes on retaining walls

As with the phenomenon of liquefaction, the earthquake behaviour of retaining walls has been extensively studied using the centrifuge. Early studies used dry sand backfill to study the nature of peak dynamic pressures and residual loads on non-failing walls, and to observe the Coulomb-type wedge failure behind sliding gravity structures. The instrumentation on these models was simple, and comprised strain gauges mounted on the walls to measure bending strains, together with accelerometers and displacement transducers distributed throughout the model. Some models tested at Cambridge University used reinforced microconcrete to study the combined problem of yeilding soil and a failing wall, but if detailed data on dynamic strains in the

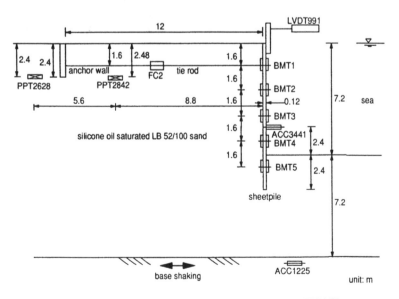

Figure 7.11 Cross-section of centrifuge model for test ZENG8.

walls themselves was necessary then steel or dural walls would be constructed, scaled to provide the bending stiffness appropriate to a concrete prototype (although this might lead to some error in scaling wall inertia and certainly would not correctly scale the plastic moment capacity). Gravity structures, where gross deformations were expected, would often use coloured sand marker bands to show up failure surfaces after the event, as seen in the sliding failure of a typical block wall retaining dry sand backfill in Figure 7.6, from Steedman (1984).

More recently centrifuge tests on waterfront retaining walls have provided important new insights into the behaviour of walls retaining saturated fill. For example data from centrifuge tests on model anchored sheet-pile walls from Cambridge University has shown that excess pore-pressure build-up during earthquakes can significantly degrade the stiffness of a retaining wall and may lead to increased amplification of base shaking (Steedman and Zeng, 1990; Zeng and Steedman, 1993). A typical centrifuge model, as shown in Figure 7.11, was instrumented with pore-pressure transducers, strain gauges and accelerometers, providing very detailed information on the dynamic behaviour of the soil–structure system. In Figure 7.12 it is clear that at the beginning of the earthquake the structure was stiff and the vibration of the wall remained in phase with the input motion. However as the pore pressure built up in the soil, the stiffness of the system deteriorated, as indicated by the phase shift between the top of the wall and the input motion, which gradually increased to 180°. Resonance in the structure then led to large cyclic bending moments in the wall and anchor forces in the tie rods. This phenomenon was also observed in centrifuge tests on anchored cantilever walls conducted by Schran et al. (1992).

The opportunities provided by this class of modelling for design studies are now starting to become well established. For example, in the recent development of the earthquake design approach for the massive Victorian concrete dock walls at Devonport, in Plymouth, England, centrifuge model test data played a key role in supporting numerical modelling and analytical solutions, providing clarification on the potential mechanisms of failure, the nature of hydrodynamic pressures and the forces that might be imposed in rock anchorages.

7.4 Blast models

7.4.1 Containment

The detonation of explosive charges within a centrifuge model chamber, whether to generate earthquake-type elastic wave energy or to model high explosions, poses particular problems for the designer. Typical charges, which may be only of the order of a few grams of TNT, could nevertheless rupture a model chamber or transmit unacceptably high dynamic stresses into the

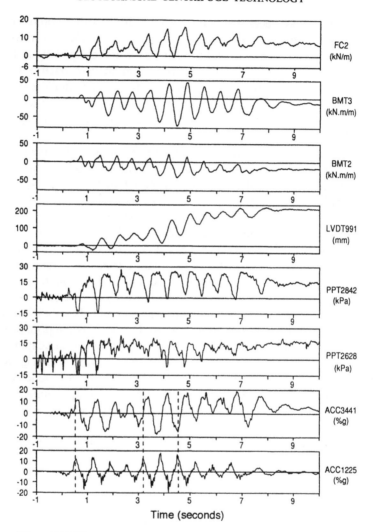

Figure 7.12 Time records of transducer outputs during EQ6, test ZENG8.

platform and (hence the centrifuge) unless the container design is adequate. One approach has been to ensure that the container is failsafe, by allowing for the venting of pressure prior to any rupture condition. Alternatively, secondary containment can be provided in the form of an inner and outer chamber, separated by an air gap.

7.4.2 Instrumentation

In many respects the instrumentation used for earthquake modelling, such as accelerometers and pore-pressure transducers, is very similar to that used in

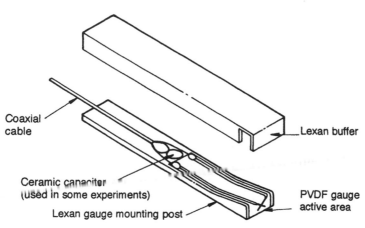

Figure 7.13 A PVDF total stress gauge for dynamic pressure measurement.

blast experiments. The range of acceleration which these accelerometers must cope with is very different, however, and care must also be taken in the use of pore-pressure transducers. Devices which are sensitive enough to measure the pore pressures in liquefied ground, for example, are unlikely to survive the high pulse of the initial elastic shock wave. Strain gauges mounted on structural elements can provide valuable information prior to yielding and passive markers in the soil can also be very effective, as can be seen in the example below.

A device which is unique to dynamic testing is the PVDF total stress gauge, which has been used successfully to measure dynamic pressure directly. An example is shown in sketch form in Figure 7.13. The transducer comprises polyvinylidene fluoride film only a few micrometres in thickness emplaced in a plastic gauge, shaped in a long rectangular probe design with the sensitive element at one end (Gaffney *et al.*, 1989). A charge is generated by the transducer when subjected to the incident stress wave and the probe design ensures the gauge is restrained by the surrounding soil long enough for the incident stress wave to be captured. Once the shock wave has enveloped the device and the gauge is moving with the soil, the stress change picked up by the gauge will not reflect the free field pressure accurately.

The apparently ever-increasing capabilities of computers and fast data-logging cards has now reached the stage that high speed data aquisition is available at reasonable cost, at least over a few channels. Whereas for earthquake modelling sampling frequencies of the order of several kilohertz are likely to be quite adequate, for blast modelling sampling in the megahertz range is desirable. Special systems utilising on-board data aquisition cards are now under development, with on-board computers controlled via remote terminals. In the future, increasing use will be made of optical joints for the passage of large volumes of digital data either in real time or immediately following an event. High-speed cameras have also been used on board to record the impact of projectiles or crater formation, sometimes filming through a mirror to observe the model in cross-section. These systems are very expensive and likely only to be justified on special projects.

7.4.3 Blast modelling of pile foundations

Much of the field data of cratering which is in the public domain, and which is available for comparison with centrifuge model data, is based on experiments in dry soils. Centrifuge tests which have been carried out, for example by Schmidt and Housen (1987) or Serrano *et al.* (1988) have generally also concentrated on impact or explosion cratering in dry sand. This work has shown clearly the value of the centrifuge in replicating field data and in particular in exploring the transition from low yield events, where the crater volume is dominated by the strength of the soil, to high yield events in which crater volume is dominated by gravity.

In the case of saturated soils there is considerably less centrifuge test data but tests which have been carried out show the importance of the depth of the water-table in determining crater shape and volume. Figure 7.14 shows a crater formed at $60\,g$ by a buried blast equivalent to 1000 lb of high explosive. The position of the water table is clearly seen at the boundary between the

Figure 7.14 Crater formation from a buried blast at $60\,g$ in saturated sand.

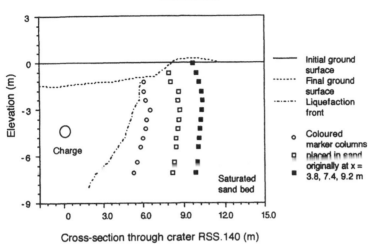

Figure 7.15 Cross-section through crater formed at 60 g.

apparent crater, infilled by blast-induced liquefaction, and the cliff-edge around the crater perimeter. In cross-section in a similar model, Figure 7.15, passive coloured markers in the sand help to define the true crater and extent of plastic strain in the surrounding soil.

This experiment was one of a series undertaken to determine the feasibility of using small scale models on a centrifuge to acquire reliable and accurate data on the blast response of piled foundations, reported by Steedman *et al.* (1989) and Gaffney *et al.* (1989). Accordingly individual piles, or a line of piles, were fixed at ground level using gantries at a stand-off distance which would place them at, or near, the edge of the crater. The piles were instrumented with strain gauges to measure bending strains, which could be interpreted as moments provided the piles were not yielding at that location.

Time histories of the bending moments on the pile at different depths (Figure 7.16) show clearly the passage of a plastic soil wave which lasts for some hundreds of milliseconds. (The single spike at time 0 was caused by the detonation of the charge.) By using a series of gauges along the length of the pile 'snapshots' of the development of bending moment with time were seen as a function of depth (Figure 7.17) during the arrival of the blast wave. The full plastic moment capacity of the pile is quickly mobilised in a hogging direction under the pile cap and later, in the opposite sense, at mid-depth.

Clearly the opportunities provided by such models for studying blast loading effects on structures surrounded by soil are considerable, and it is to be expected that with the development of a major new centrifuge facility by the US Army Corps of Engineers Waterways Experiment Station at Vicksburg in 1995, more research of this type will be carried out in future.

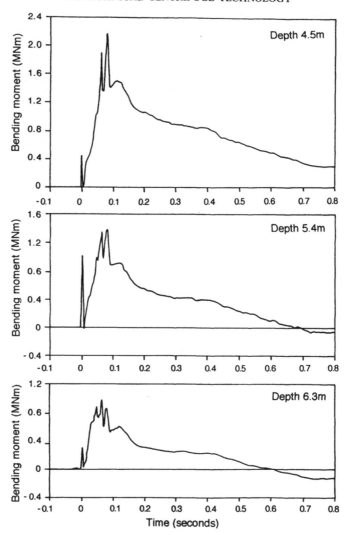

Figure 7.16 Build-up of bending moment in pile under blast loading.

7.5 Dynamic modelling

Dynamic modelling will continue to occupy an important niche in the range of applications within geotechnical centrifuge modelling. Recently, the design of containment systems for earthquake models has advanced considerably. In the future more sophisticated actuators will become available as advances are made in the design of mechanical valves and costs reduce. The opportunities for clarifying our understanding of the earthquake safety of structures, dams and lifelines are only beginning to be tapped; this will have important benefits for designers and risk analysts as the consequences of ground failure in earthquakes are modelled more accurately.

The study of weapons effects using centrifuges is likely to increase as financial pressures on military budgets and environmental pressures on field

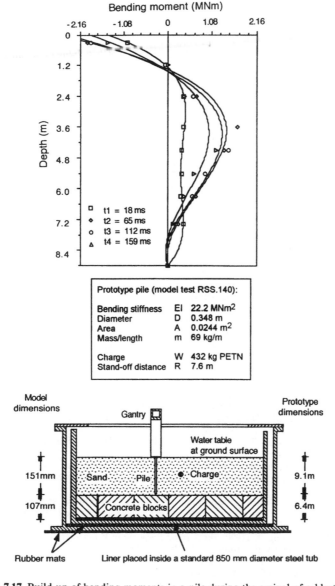

Figure 7.17 Build-up of bending moments in a pile during the arrival of a blast wave.

testing grow stronger. Blast modelling will remain one of the key fields which is ideally served by a high gravity capability. Although many experiments will continue to be conducted very successfully at modest accelerations, the scaling advantage of high gravity work, in terms both of size of charge and depth and area of site, will lead to an increase in the standard operating g levels of large centrifuges.

References

Committee on Earthquake Engineering (1985) *Liquefaction of Soils During Earthquakes*. Report of the National Research Council, National Academy Press, Washington, D.C.

Gaffney, E.S., Felice, C.W. and Steedman, R.S. (1989) Comparison of cratering in wet media by buried charges: centrifuge and field events. *4th Int. Symp. Interaction of Non-nuclear Munitions with Structures*, April 17–21, Vol. 1, pp. 402–407. Panama City Beach, Fl.

Hushmand, B., Scott, R.F. and Crouse, C.B. (1988) Centrifuge liquefaction tests in a laminar box. *Géotechnique*, **2**, 253–262.

Iai, S. (1989) Similitude for shaking table tests on soil–structure-fluid model in 1 g gravitational field. *Soils Found. Japan Soc. Soil Mech. Found. Eng.*, **29** (1), 105–118.

Ketcham, S.A., Ko, H.-Y. and Sture, S. (1988) An electrohydraulic earthquake simulator for centrifuge testing. *Centrifuge '88*, pp. 97–102. Balkema, Rotterdam.

Kutter, B.L. (1982) Centrifuge Modelling of the Response of Clay Embankments to Earthquakes. PhD Thesis, Cambridge University.

Mikasa, M., Takada, N. and Yamada, K. (1969) Centrifugal model test of a rockfill dam. *Proc. 7th Int. Conf. Soil Mech. Found. Eng., Mexico*, Vol. 2, pp. 325–333, Sociedad Mexicana de Mecanica de Suelos, AC.

Schmidt, R.M. and Housen, K.R. (1987) Some recent advances in the scaling of impact and explosion cratering, *Int. J. Impact Eng.*, **5**, 543–560.

Schofield, A.N. (1981) Dynamic and earthquake geotechnical centrifuge modelling. *Proc. Int. Conf. Recent Advances in Geotechnical Earthquake Engineering and Soil Dynamics*, Vol. 3, pp. 1081–1100. University of Missouri-Rolla, Rolla, MO.

Schofield, A.N. and Steedman, R.S. (1988) Recent development of dynamic model testing in geotechnical engineering. *Proc. 9th World Conf. Earthquake Eng.*, Japan Association for Earthquake Disaster Prevention, August 2–9, Kyoto-Tokyo, Vol. 8, pp. 813–824.

Schofield, A.N. and Zeng, X. (1992) *Design and Performance of an Equivalent Shear Beam (ESB) Container for Earthquake Centrifuge Modelling*, Technical Report CUED/D-SOILS/TR245, Department of Engineering, Cambridge University.

Schran, UK., Ting, N.H. and Whitman, R.V. (1992) *Dynamic Centrifuge Testing of a Tilting Retaining Wall with Saturated Backfill*. Technical Report. Department of Civil Engineering, Massachusetts Institute of Technology, Boston, MA.

Serrano, C.H., Dick, R.D. Goodings, D.J. and Fourney, W.L. (1988) Centrifuge modelling of explosion induced craters. *Centrifuge '88, Paris*, pp. 445–450. Balkema, Rotterdam.

Steedman, R.S. (1984) Modelling the Behaviour of Retaining Walls in Earthquakes. PhD Thesis, Cambridge University.

Steedman, R.S. (1990) Discussion on similitude for shaking table tests on soil–structure-fluid model in 1 g gravitational field by S. Iai. *Soils Found. J. Japan Soc. Soil Mech. Found. Eng.*, **30** (2), June.

Steedman, R.S. (1991) Centrifuge modelling for dynamic geotechnical studies. *2nd Int. Conf. Recent Advances in Geotechnical Earthquake Engineering and Soil Dynamics*, Vol. 3, pp. 2401–2417, St. Louis, MO.

Steedman, R.S. and Zeng, X. (1990) *The Seismic Response of Waterfront Retaining Walls*. ASCE Geotechnical Special Publication No. 25, Ithaca, NY.

Steedman, R.S., Felice, C.W. and Gaffney, E.S. (1989) Dynamic response of deep foundations. *4th Int. Symp. Interaction of Non-nuclear Munitions with Structures*, April 17–21, Vol. 2, pp. 80–84. Panama City Beach, FL.

Whitman, R.V. (1988) Experiments with earthquake ground motion simulation. In *Centrifuges in Soil Mechanics*, (ed. W.H. Craig *et al.*), pp. 203–216. Balkema, Rotterdam.

Whitman, R.V., Lambe, P.C. and Kutter, B.L. (1981) Initial results from a stacked ring apparatus for simulation of a soil profile. *Proc. Int. Conf. on Recent Advances in Geotechnical Earthquake Engineering and Soil Dynamics*, Vol. 1, pp. 361–366, University of Missouri-Rolla, Rolla, MO.

Zeng, X. and Steedman, R.S. (1993) On the behaviour of quay walls in earthquakes. *Géotechnique*, **43** (3), 417–431.

8 Environmental geomechanics and transport processes

P.J. CULLIGAN-HENSLEY and C. SAVVIDOU

Notation

a_f	coefficient relating fluid density to contaminant mass fraction
a_L	longitudinal dispersion coefficient
a_w	coefficient of volume expansion of fluid
c	volume-based contaminant concentration in the fluid phase
c_f	heat capacity of the fluid phase at constant pressure
c_o	concentration of injected contaminant species
c_s	heat capacity of the solid phase at constant pressure
c_s	concentration of model pollutant (Figure 8.6)
C_a	capillary effects number
d	characteristic microscopic length (e.g. particle size)
$\mathbf{D}$	mechanical dispersion tensor
D	mechanical dispersion coefficient
D_d	free diffusion coefficient of contaminant in solution
D_d^*	effective diffusion coefficient of contaminant in soil
$\mathbf{D_h}$	hydrodynamic dispersion tensor $(\mathbf{D_h} = D_d^*\mathbf{I} + \mathbf{D})$
D_h	hydrodynamic dispersion coefficient $(D_h = D_d^* + D)$
g	gravitational constant
h_c	capillary head
$\mathbf{I}$	identity matrix
I	inter-region, contaminant transport number
$\mathbf{k}$	intrinsic permeability tensor
k	intrinsic permeability coefficient
K	hydraulic conductivity
K_d	equilibrium distribution coefficient
K_f	thermal conductivity of the fluid phase
K_s	thermal conductivity of the solid phase
l_c	characteristic length
n	effective porosity
m	mean
N	scaling factor
p	fluid pressure
P_e	Peclet number

P_r	Prandtl number
$\mathbf{q}$	specific discharge vector
$\mathbf{q}_r$	specific discharge vector relative to the (possibly) moving solid
Q	outflow
Q_{inf}	outflow at infinite time
R_{ae}	effective Rayleigh number
R_{as}	solute Rayleigh number
R_{aT}	thermal Rayleigh number
R_e	Reynolds number
R_{fs}	transfer rate of solute from the fluid to the solid phase per unit mass of solid phase
s	standard deviation
S	Schmidt number
t	time
T	porous medium temperature
T'	tortuosity of porous medium
$\mathbf{u}$	interstitial fluid velocity vector
u	one-dimensional interstitial fluid velocity
$\mathbf{u}_s$	average velocity vector of solid grains
u_s	average one-dimensional velocity of solid grains
w	mass fraction of contaminant in the fluid phase
w_s	mass fraction of contaminant on the solid phase
z	vertical dimension
z'	position of observer moving with the average flow
Z	typical linear dimension
α	inter-region contaminant mass transfer rate
ε_v	volumetric strain
θ	volumetric fluid content
κ	thermal diffusivity of porous medium
λ	linear decay rate constant
μ	fluid viscosity
ν	kinematic fluid viscosity
ρ	fluid density
ρ_b	bulk density of the porous medium
ρ_s	density of the solid phase
σ	surface (or interfacial) tension
∇	del operator
∂	partial operator

Subscripts

m	model
p	prototype
r	ratio of value in the prototype to that in the model

8.1 Introduction

The objective of any modelling exercise is one of simulation and prediction. In the particular case of sub-surface transport processes, modelling can provide valuable information about the movement of groundwater, the spread and growth of heat and pollutant plumes, and the effectiveness of various containment and remedial action strategies. Thus, modelling can offer much assistance in identifying optimum strategies for sound waste management.

Predicting the movement of heat and contaminants in groundwater flow systems is often carried out using theoretical modelling techniques. During theoretical modelling, the processes under examination are simulated by a set of governing equations, that are subsequently solved using either analytical or numerical methods. The results of any theoretical modelling exercise are wholly dependent upon complete understanding of the fundamental processes involved, and accurate conceptual modelling of all relevant mechanisms.

In recent years, it has become apparent that a critical need exists for physical observations of transport mechanisms in soils. Such observations are required both to validate existing mathematical models, and to aid in developing improved conceptual models of fundamental processes.

Historically, controlled field experiments and laboratory column tests have provided the bulk of experimental data on the fate of heat and/or chemicals being transported in soil. Controlled field experiments have the advantage of modelling the total complexity of the full-scale problem. However, these tests are costly, difficult to perform and often offer little direct control over boundary conditions. On the other hand, laboratory column experiments, which are generally inexpensive and relatively uncomplicated to perform, are often of limited value due to their inability to model realistic prototype conditions.

Recently, researchers have come to recognise that a geotechnical centrifuge can provide a powerful experimental tool for investigation of many environmental engineering problems. A geotechnical centrifuge has the ability to model complex two- and three-dimensional problems, under repeatable and controlled boundary conditions.

This chapter discusses the fundamental transport mechanisms involved in many environmental engineering problems, and outlines the principles and scaling laws related to the centrifuge modelling of such mechanisms. A review of centrifuge modelling work carried out in the area of environmental geotechnics is then presented. The scope of this work serves to demonstrate the significant contribution that centrifuge modelling can make to the field of Environmental Engineering.

8.2 Transport mechanisms

Many geo-environmental problems concern the transport of fluid mass, energy (in the case of heat transport), and/or contaminant mass, through a porous medium. In general, the transport of any quantity through a porous medium is classically described by a conservation equation that models the fate of that quantity in a macroscopic continuum.

This section presents fundamental conservation equations that describe mass, heat and contaminant transport in a porous medium. For simplicity, only the general case of saturated flow in a deformable medium is considered here. However, more specific cases are discussed in a later section. A full derivation of all fundamental equations can be found in Bear (1972).

8.2.1 Mass transport

The general equation describing mass transport of a single fluid phase in a saturated, deformable matrix, is based upon conservation of mass in a volume element coupled with an equation of fluid motion.

In the absence of any external fluid sources, and neglecting diffusive and dispersive fluxes of fluid mass, the macroscopic balance equation for fluid mass is given by[1]:

$$\nabla \cdot (\rho \mathbf{q}) + \frac{\partial (n\rho)}{\partial t} = 0 \qquad (8.1)$$

where ρ is the fluid density, n is the effective porosity and $\mathbf{q}$ is the specific discharge (volume of fluid flow per unit time per unit area).

For a deformable matrix, where solid grains are moving at an average velocity $\mathbf{u_s}$ with respect to fixed coordinates,

$$\mathbf{q} = n\mathbf{u} = \mathbf{q_r} + n\mathbf{u_s} \qquad (8.2)$$

where $\mathbf{u}$ is the average pore fluid velocity, $\mathbf{q_r}$ is the specific discharge with respect to the moving solid particles and n is the instantaneous porosity.

By combining equations (8.1) and (8.2) we obtain:

$$\nabla \cdot (\rho \mathbf{q_r}) + n\mathbf{u_s} \cdot \nabla \rho + \rho \mathbf{u_s} \cdot \nabla n + \rho n \cdot \nabla \mathbf{u_s} + n \frac{\partial \rho}{\partial t} + \rho \frac{\partial n}{\partial t} = 0 \qquad (8.3)$$

The mass balance equation for the solid phase can be expressed by an equation similar to equation (8.1), viz.

$$\nabla \cdot [(1-n)\rho_s \mathbf{u_s}] + \frac{\partial (1-n)\rho_s}{\partial t} = 0 \qquad (8.4)$$

where ρ_s is the density of the solid.

If we assume that the macroscopic density of the solid remains unchanged

---

[1] Unless otherwise stated, an orthogonal, right-handed coordinate system has been adopted with the z-axis pointing vertically upward: it is further assumed that the coordinate system is aligned with the principal directions of the permeability tensor.

while the medium as a whole is undergoing deformation, equation (8.4) reduces to:

$$\nabla \cdot \mathbf{u_s} = -\frac{1}{(1-n)}\left[\frac{\partial(1-n)}{\partial t} + \mathbf{u_s} \cdot \nabla(1-n)\right] \qquad (8.5)$$

By eliminating $\nabla \cdot \mathbf{u_s}$ from equations (8.3) and (8.5), and by further assuming that $|\partial(..)/\partial t| \gg |\mathbf{u_s} \cdot \nabla(..)|$ in all cases, we obtain:

$$\nabla \cdot (\rho \mathbf{q_r}) + n\frac{\partial \rho}{\partial t} + \rho\left(\frac{1}{(1-n)}\frac{\partial n}{\partial t}\right) = 0 \qquad (8.6)$$

If we recall that $[1/(1-n)]\partial n/\partial t \equiv [1/(1+e)]\partial e/\partial t$, where e is the instantaneous void ratio of the medium, equation (8.6) can also be written in the form:

$$\nabla \cdot (\rho \mathbf{q_r}) + n\frac{\partial \rho}{\partial t} + \rho\frac{\partial \varepsilon_v}{\partial t} = 0 \qquad (8.7)$$

where ε_v is the volumetric strain (or dilation) of the porous medium ($\equiv \partial e/(1+e)$).

For laminar flow, the specific discharge with respect to the (possibly) moving solid particles is given by Darcy's law. Thus

$$\mathbf{q_r} = -\frac{\mathbf{k}}{\mu}(\nabla p + \rho g \nabla z) \qquad (8.8)$$

where $\mathbf{k}$ is the intrinsic permeability of the porous medium, μ is the fluid viscosity, p is the fluid pressure and g is the gravitational constant.

A general mass conservation equation for the fluid is obtained by combining equations (8.6) and (8.8). Hence:

$$-\nabla \cdot \left[\rho\frac{\mathbf{k}}{\mu}(\nabla p + \rho g \nabla z)\right] + n\frac{\partial \rho}{\partial t} + \rho\left(\frac{1}{1-n}\frac{\partial n}{\partial t}\right) = 0 \qquad (8.9)$$

8.2.2   Heat transport

Heat transport through porous media can take place through pure conduction alone, in which case heat is transferred through 'static' pore fluid and through the solid particles themselves, or it can take place by convection, in which case heat is predominately transferred by the physical motion of the pore fluid itself.

Heat transport is described by a thermal energy balance equation that is based upon conservation of enthalpy in the fluid and solid phases of the porous medium.

In the absence of any external heat or fluid sources, and neglecting secondary effects such as heating due to viscous dissipation, the Dufour effect and thermal dispersion, the energy balance equation for a unit volume of fluid and solid phase together is given by:

$$\frac{\partial}{\partial t}(n\rho c_f + (1-n)\rho_s c_s)T = \nabla \cdot (nK_f + (1-n)K_s)\mathbf{I}\nabla T - \nabla \cdot \rho c_f(\mathbf{u}-\mathbf{u}_s)T$$

$$(8.10)$$

where T is the porous medium temperature, c_f is the heat capacity of the fluid phase at constant pressure, c_s is the heat capacity of the solid phase at constant pressure, K_f is the thermal conductivity of the fluid phase, K_s is the thermal conductivity of the solid phase and $\mathbf{I}$ is the identity matrix.

Equation (8.10) relates the rate of change of fluid and solid phase enthalpy to the net conductive enthalpy flux and the net advective enthalpy flux.

For purely conductive heat transfer, the temperature field is uncoupled from the flow field, and the net advective enthalpy flux is zero. In this case, heat transport is governed by Fourier's law:

$$\frac{\partial T}{\partial t} = \nabla \cdot (\kappa \nabla T) \qquad (8.11)$$

where κ is the thermal diffusivity of the medium:

$$\kappa = \frac{nK_f + (1-n)K_s}{n\rho c_f + (1-n)\rho_s c_s} \qquad (8.12)$$

8.2.3 Contaminant transport

A mass balance equation describing the transport of a soluble contaminant can be derived by considering the conservation of a single species in a unit volume of porous medium. The dependent variable for contaminant transport is taken as the mass fraction of contaminant in the fluid phase, i.e. the amount of contaminant per unit mass of fluid.

In the absence of any external sources of contaminant mass, and neglecting contaminant mass flux due to the Soret effect, the balance equation for a soluble contaminant in the fluid phase of a porous medium is given by:

$$\frac{\partial(n\rho w)}{\partial t} = \nabla \cdot n\mathbf{D}\nabla w + \nabla \cdot n\mathbf{D}_d^*\mathbf{I}\nabla w - \nabla \cdot n\mathbf{u}w - \lambda nw - \rho_b R_{fs} \qquad (8.13)$$

where w is the mass fraction of contaminant in the fluid phase, $\mathbf{D}$ is the mechanical dispersion coefficient tensor, $\mathbf{D}_d^*$ is the effective molecular diffusion coefficient of the contaminant within the medium, λ is a linear decay rate constant that describes the rate of internal decay of contaminant mass in solution due to processes such as, for example, biodegradation, ρ_b is the bulk density of the porous medium and R_{fs} is the transfer rate of contaminant from fluid to the solid phase per unit mass of solid phase.

The balance equation for conservation of contaminant mass in the solid phase is described by:

$$\frac{\partial(\rho_b w_s)}{\partial t} = \rho_b R_{fs} - \lambda \rho_b w_s \tag{8.14}$$

where w_s is the mass fraction of contaminant on the solid phase.

If we assume that the only contaminant–solid phase interaction is linear equilibrium sorption, then the mass fraction of contaminant on the solid phase is given by:

$$w_s = K_d w \tag{8.15}$$

where K_d is an equilibrium distribution coefficient.

The contaminant mass balance equation for a unit volume of fluid and solid phase together is obtained by combining equations (8.13), (8.14) and (8.15). Thus:

$$\frac{\partial}{\partial t}(n + \rho_b K_d)\rho w = \nabla \cdot n[\mathbf{D} + D_d^* \mathbf{I}]\nabla w - \nabla \cdot n\mathbf{u}w - \lambda(n + \rho_b K_d)w \tag{8.16}$$

Equation (8.16) relates the rate of change of contaminant in the fluid phase to the net dispersive and diffusive flux, the net advective flux and an internal solute source and decay rate.

Note that under conditions where the fluid density remains essentially invariant, the volume based concentration[2] c ($\equiv \rho w$) is more frequently taken as the dependent variable for contaminant transport.

For any one problem, the balance equations (8.9), (8.10) and (8.16) may be linked to each other through the (possible) dependence of fluid density, fluid viscosity and matrix porosity on fluid pressure, medium temperature and contaminant mass fraction.

8.3 Fundamentals of modelling

8.3.1 Scaling laws

For the correct modelling of a problem, it is necessary to simulate the behaviour of the prototype, but preferably on a miniature dimensional scale and at an accelerated time scale. For similitude, the characteristic equations in both systems must represent the same principles that govern the problem under investigation. Here, a derivation is presented of the scaling necessary in the centrifuge modelling of mass, heat and contaminant transport through a saturated, deformable medium. Subscript (p) denotes parameters in the prototype, subscript (m) denotes parameters in the centrifuge model and

---

[2] That is, the amount of contaminant per unit volume of fluid.

subscript (r) denotes the ratio between a value in the prototype to that in the model. To simplify discussion, the scaling laws will be derived using one-dimensional expressions. Nevertheless, the obtained laws will be equally applicable for multi-dimensional problems.

For a centrifuge model having vertical dimensions z_m and tested at an acceleration of N gravities, the following scaling ratios are evident:

$$z_r = \frac{z_p}{z_m} = N \qquad (8.17)$$

and

$$g_r = \frac{g_p}{g_m} = \frac{1}{N} \qquad (8.18)$$

In the ensuing paragraphs, scaling ratios will be derived for the dependent variables, fluid pressure (p), medium temperature (T) and contaminant mass fraction (w), for the (possibly) variant fluid and matrix properties, density ($\rho = f(p, T, w)$), viscosity ($\mu = f(T, w)$) and porosity ($n = f(p, T)$), for the fluid and solid grain velocities (u and u_s), and for the time scale (t). The derivation of these ratios will be done by examining scaling of the governing equations for mass, heat and contaminant transport. It will be assumed that (i) the centrifuge model tests are conducted using prototype soil and fluids, (ii) the temperature of a heat source is identical in model and prototype and (iii) the concentration of a pollutant source is identical in model and prototype. Furthermore, it will be assumed that the material parameters k, c_f, c_s, ρ_s, D_d^*, K_f, K_s and K_d are identical at homologous points in model and prototype.[3] Note that the mechanical dispersion coefficient D is a function of both interstitial velocity and space, and thus cannot be considered a material parameter.

8.3.1.1 Mass transport.

The general equation of motion for mass transport was given by equation (8.9). This equation can be written in one-dimensional form for the prototype as:

$$-\frac{\partial}{\partial z_p}\left[\rho_p \frac{k}{\mu_p}\left(\frac{\partial p_p}{\partial z_p} + \rho_p g_p\right)\right] + n_p \frac{\partial \rho_p}{\partial t_p} + \rho_p\left(\frac{1}{(1-n_p)}\frac{\partial n_p}{\partial t_p}\right) = 0 \qquad (8.19)$$

and for the model as:

$$-\frac{\partial}{\partial z_m}\left[\rho_m \frac{k}{\mu_m}\left(\frac{\partial p_m}{\partial z_m} + \rho_m g_m\right)\right] + n_m \frac{\partial \rho_m}{\partial t_m} + \rho_m\left(\frac{1}{(1-n_m)}\frac{\partial n_m}{\partial t_m}\right) = 0$$

$$(8.20)$$

---

[3] In some cases, these material parameters may be functions of fluid pressure, medium temperature and/or matrix porosity. It is thus assumed that if similitude of pressure, temperature and porosity is obtained between model and prototype, similitude between these (instantaneous) parameters will also be achieved.

Equation (8.19), referring to the prototype, can be re-written as:

$$-\frac{1}{z_r}\frac{\partial}{\partial z_m}\left[\rho_r\rho_m\frac{1}{\mu_r}\frac{k}{\mu_m}\left(\frac{p_r}{z_r}\frac{\partial p_m}{\partial z_m}+\rho_r\rho_m g_r g_m\right)\right]+n_r n_m\frac{\rho_r}{t_r}\frac{\partial \rho_m}{\partial t_m}$$

$$+\rho_r\rho_m\left(\frac{1}{(1-n_r n_m)}\frac{n_r}{t_r}\frac{\partial n_m}{\partial t_m}\right)=0 \quad (8.21)$$

By comparing equation (8.20) with equation (8.21), we find, that for correct scaling:

$$\frac{\rho_r}{\mu_r}\frac{p_r}{z_r^2}\equiv\frac{\rho_r^2}{\mu_r}\frac{g_r}{z_r}\equiv n_r\frac{\rho_r}{t_r}\equiv\rho_r\frac{n_r}{t_r} \quad (8.22)$$

8.3.1.2 Heat transport.

The energy balance equation for heat transport was given by equation (8.10). By following the above procedure, the energy balance equation for the prototype can be written as:

$$\frac{1}{t_r}\frac{\partial}{\partial t_m}[(n_r n_m\rho_r\rho_m c_f+(1-n_r n_m)\rho_s c_s)T_r T_m]=\frac{1}{z_r}\frac{\partial}{\partial z_m}$$

$$\times\left((n_r n_m K_f+(1-n_r n_m)K_s)\frac{T_r}{z_r}\frac{\partial T_m}{\partial z_m}\right)-\frac{1}{z_r}\frac{\partial}{\partial z_m}(n_r n_m\rho_r\rho_m c_f(u_r u_m-u_{s_r} u_{s_m})T_r T_m)$$

$$(8.23)$$

By comparing this with the energy balance equation for heat transport in the model:

$$\frac{\partial}{\partial t_m}[(n_m\rho_m c_f+(1-n_m)\rho_s c_s)T_m]=\frac{\partial}{\partial z_m}\left((n_m K_f+(1-n_m)K_s)\frac{\partial T_m}{\partial z_m}\right)$$

$$-\frac{\partial}{\partial z_m}(n_m\rho_m c_f(u_m-u_{s_m})T_m) \quad (8.24)$$

we find that the following conditions must hold for correct scaling:

$$n_r\rho_r\frac{T_r}{t_r}\equiv n_r\frac{T_r}{t_r}\equiv n_r\frac{T_r}{z_r^2}\equiv n_r\rho_r u_r\frac{T_r}{z_r}\equiv n_r\rho_r u_{s_r}\frac{T_r}{z_r} \quad (8.25)$$

8.3.1.3 Contaminant transport.

The mass balance equation for contaminant transport was given by equation (8.16). This equation may be written for the prototype as:

$$\frac{1}{t_r}\frac{\partial}{\partial t_m}[(n_r n_m+\rho_b K_d)\rho_r\rho_m w_r w_m]=\frac{1}{z_r}\frac{\partial}{\partial z_m}[n_r n_m\rho_r\rho_m(D_r D_m+D_d^*)]\frac{w_r}{z_r}\frac{\partial w_m}{\partial z_r}$$

$$-\frac{1}{z_r}\frac{\partial}{\partial z_m}(n_r n_m\rho_r\rho_m u_r u_m w_r w_m)-\lambda_r\lambda_m(n_r n_m+\rho_b K_d)\rho_r\rho_m w_r w_m \quad (8.26)$$

and for the model as:

$$\frac{\partial}{\partial t_m}[(n_m + \rho_b K_d)\rho_m w_m] = \frac{\partial}{\partial z_m}\left(n_m \rho_m [D_m + D_d^*]\frac{\partial w_m}{\partial z_m} - \frac{\partial}{\partial z_m}(n_m \rho_m u_m w_m)\right.$$

$$\left. - \lambda_m (n_m + \rho_b K_d)\rho_m w_m\right) \quad (8.27)$$

By comparing equations (8.26) and (8.27), it is clear that the following relationship must hold for correct scaling:

$$n_r \rho_r \frac{w_r}{t_r} \equiv n_r \rho_r D_r \frac{w_r}{z_r^2} \equiv n_r \rho_r \frac{w_r}{z_r^2} \equiv n_r \rho_r u_r \frac{w_r}{z_r} \equiv \lambda_r n_r \rho_r w_r \equiv \lambda_r \rho_r w_r \quad (8.28)$$

Taking into account equations (8.17) and (8.18), it is straightforward to show that equations (8.22), (8.25) and (8.28) are satisfied when:

$$p_r = T_r = w_r = 1$$
$$\rho_r = \mu_r = n_r = 1$$
$$u_r = u_{s_r} = \frac{1}{N} \quad (8.29)$$
$$t_r = N^2$$
$$D_r = 1; \quad \lambda_r = \frac{1}{N^2}$$

8.3.1.4 Summary. Discussion on the scaling laws has shown that transport events will occur N^2 times faster in a reduced scale centrifuge model than the full-scale prototype. The discussion also disclosed that pressure, temperature and concentration changes will be identical at homologous points in the model and prototype, and further revealed that the flow and solid grain velocities in the centrifuge model will be N times higher than the corresponding velocities in the prototype.

A summary of the scaling laws is given in Table 8.1. The validity of these

Table 8.1 Centrifuge scaling factors for correct modelling

Parameter	Symbol	Prototype–model ratio
Gravity	g	$1/N$
Length	z	N
Pressure	p	1
Temperature	T	1
Contaminant mass fraction	w	1
Density	ρ	1
Viscosity	μ	1
Porosity	n	1
Velocity	u	$1/N$
Time	t	N^2
Mechanical dispersion coefficient	D	1
Linear decay rate constant	λ	$1/N^2$

laws under certain conditions has been demonstrated by Arulanandan *et al.* (1988), Savvidou (1988) and Celorie *et al.* (1989).

A discussion on the validity of scaling is given in section 8.3.3.

8.3.2 Dimensionless groups

The significance of dimensionless groups relevant to the centrifuge modelling of mass, heat and contaminant transport is discussed here. For brevity, the discussion includes only those groups pertinent to the work reviewed in this chapter. Consequently, the following catalogue is by no means exhaustive.

For strict similitude between centrifuge model and prototype, the groups listed below should be identical in model and prototype. However, in certain cases, the same principles of behaviour will govern the model and prototype even when similitude is not achieved. Further discussion on this point is given in section 8.3.3.

8.3.2.1 Reynolds number. The Reynolds number (R_e) describes the ratio between inertial and viscous forces in a fluid. For fluid flow through a porous medium, its value is defined by:

$$R_e = \frac{\rho u d}{\mu} \tag{8.30}$$

where d represents a characteristic microscopic length of the medium, such as soil particle size.

The value of the Reynolds number categorises the manner by which fluid flows through a medium (Bear, 1972). At low Reynolds numbers ($R_e \leqslant 10$), viscous forces predominate, and flow is laminar. At high Reynolds numbers ($R_e \geqslant 100$) inertial forces govern, and flow is turbulent. At intermediate Reynolds numbers ($10 < R_e < 100$), a transition zone exists, where inertial forces influence laminar flow.

Exact similarity of fluid motion between model and prototype will arise if $R_{em} = R_{ep}$. However, even if $R_{em} \neq R_{ep}$, similarity of the *equation* of fluid motion between model and prototype can still arise, provided that the same manner of fluid flow occurs in both cases. This point is further discussed in section 8.3.3.

8.3.2.2 Mass transport Peclet number. The mass transport Peclet number (P_e) describes the ratio between advective mass transport and diffusive mass transport. For porous media flow, its value is described by:

$$P_e = \frac{u l_c}{D_d} \tag{8.31}$$

where l_c is some characteristic length of the system and D_d is the free diffusion coefficient of the contaminant in solution.

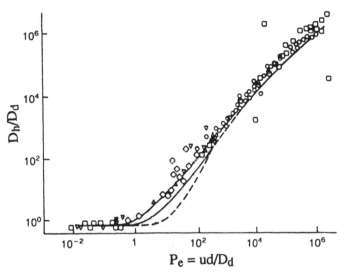

Figure 8.1 The dependence of dispersion on the Peclet number. (After Bear and Verruijt, 1987, from Hensley and Randolph, 1994.)

For problems concerning contaminant transport, the value of P_e characterises the relationship between the two dispersive (spreading) phenomena of molecular diffusion and mechanical dispersion. For tests conducted in uniform soil, much experimental and laboratory work has shown the relationship given in Figure 8.1 between the hydrodynamic dispersion coefficient D_h, which includes the effects of both mechanical dispersion and molecular diffusion $(D_h = D + D_d^*)$, and the dimensionless Peclet number $P_e = ud/D_d$, in which the characteristic length effecting spreading is described by the soil's characteristic microscopic length, d.

For the range of Peclet numbers normally encountered in practice, three regions of behaviour are identifiable (Bear, 1979). For $P_e \leqslant 0.4$, contaminant dispersion is dominated by the process of molecular diffusion. For $P_e \geqslant 5$, contaminant dispersion is dominated by mechanical dispersion. For intermediate values $(0.4 < P_e < 5)$ a transition zone exists, where dispersion is governed simultaneously by molecular diffusion and mechanical dispersion.

The process of contaminant dispersion will be similar between a model and the prototype when $(D_h/D_d)_m = (D_h/D_d)_p$. This will be the case if P_e remains below unity in both the model and prototype, or if the characteristic length affecting dispersion, l_c, is scaled between model and prototype (in other words, if $P_{em} = P_{ep}$). Further discussion on this latter point is given in section 8.3.3.2.

8.3.2.3 Rayleigh numbers. For porous media flow, the Rayleigh number describes the form of hydraulic instability generated in a medium whose interstitial fluid is stratified in density.

For problems involving heat and/or contaminant transport, stratifications in fluid density may be formed by variations in temperature and/or concentration between adjacent layers of fluid.

For the non-isothermal flow of a homogeneous fluid with $\rho = f(T)$, the onset of hydraulic instability is described by a 'thermal' Rayleigh number R_{aT}, which symbolises a balance between the driving buoyancy force induced by free convective motion and the damping processes of viscous resistance and thermal diffusivity. For linear problems, the thermal Rayleigh number is defined by (Elder, 1967)

$$R_{aT} = \frac{\left(-\frac{\partial T}{\partial z}\right) gkZ^2 a_w}{\kappa v} \qquad (8.32)$$

where $\partial T/\partial z$ is the vertical temperature gradient at any given z, a_w is the coefficient of volume expansion of the fluid, v is the kinematic viscosity of the pore fluid, and Z is a typical linear dimension of the problem.

Evidence suggests that convective heat transfer first appears when the thermal Rayleigh number is increased above a critical value $R_{aT} = 4\pi^2$ (Lapwood, 1948; Elder, 1967). Thus, below this critical value, heat transfer is by pure conduction alone. However, above this critical value, the pattern of heat transfer is governed by the magnitude of R_{aT} (Bejan, 1984).

For the isothermal flow of an inhomogeneous fluid with $\rho = f(w)$, the gravitational stability of contaminant transport is described by a 'solute' Rayleigh number, which again symbolises the balance between a driving buoyancy force and the resistive processes of viscosity and diffusivity. For linear problems, the solute Rayleigh number is described by an equation similar to equation (8.32) (Wooding, 1959), viz.

$$R_{as} = \frac{\left(\frac{\partial w}{\partial z}\right) gkZ^2 a_f}{D_d^* v} \qquad (8.33)$$

where $\partial w/\partial z$ is the vertical concentration gradient at any given z, and a_f is a parameter which relates changes in fluid density to changes in w.

Again, evidence suggests that unstable convective transport first appears when the solute Rayleigh number is increased above a critical value, $R_{as} = 4\pi^2$ (Wooding, 1959; Schowalter, 1965). Below this critical value, the miscible displacement of one fluid by another is stable. However, above this critical value, the formation of convection cells or solute fingers becomes possible.

For the non-isothermal flow of an inhomogeneous fluid with $\rho = f(T, w)$, the problem concerns density gradients of two components with different diffusivities. When these gradients have opposing effects on the vertical density distribution, the existence of a net density distribution that increases vertically upwards is no longer a guarantee of fluid stability (Turner, 1973). Diffusion, which acts as a stabilising process in a single-component system, can now act to destabilise the system.

When a parcel of fluid in a stratified, dual-component system is displaced vertically, slow transfer of contaminant concentration relative to heat is assured by the smaller molecular diffusivity of most contaminants relative to their thermal diffusivity. In the limiting case, fluid parcels will come into thermal equilibrium with the surrounding fluid whilst still maintaining their concentration.

If the concentration gradient is destabilising and the temperature gradient stabilising, diffusion of heat into or out of a displaced fluid parcel will cause the parcel to be accelerated in the direction of its initial displacement (Figure 8.2a). Monotonic instabilities thus form, because the effect of heat exchange allows contaminated parcels to sink and fresh (uncontaminated) parcels to rise.

If the concentration gradient is stabilising and the temperature gradient destabilising, diffusion of heat will reverse the direction of the displacement, and so cause overstability or oscillatory instability (Figure 8.2b).

The form of hydraulic instability generated in a medium whose interstitial fluid is stratified by both heat and solute is described by an effective Rayleigh number, R_{ae} (Neild, 1968).

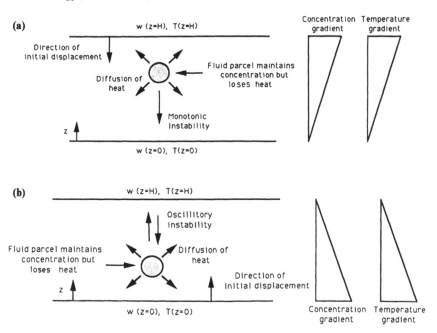

Figure 8.2 Motion of displaced fluid parcels: (a) concentration gradient destabilising and temperature gradient stabilising; (b) concentration gradient stabilising and temperature gradient destabilising. (From Hensley and Savvidou, 1993.)

For conditions of monotonic instability, linear theory indicates that R_{ae} is given by:

$$R_{ae} = R_{aT} + R_{as} \qquad (8.34)$$

while for conditions of oscillatory over-stability,

$$R_{ae} = \left(\frac{S}{P_r + S}\right) R_{aT} + \left(\frac{P_r}{P_r + S}\right) R_{as} \qquad (8.35)$$

where $P_r = v/\kappa$ and $S = v/D_d^*$ are the Prandtl and Schmidt numbers of the fluid, respectively.

As before, evidence suggests a critical value of the effective Rayleigh number $R_{ae} = 4\pi^2$, above which various forms of instability may be observed.

The form of hydraulic instability generated in a medium under the influence of density gradients will be identical in model and prototype provided that $R_{am} = R_{ap}$. For centrifuge modelling, it is straightforward to demonstrate that $R_{am} = R_{ap}$ for all of the cases presented in equations (8.32)–(8.35). Thus, the effects of fluid stratification will be correctly modelled during centrifuge testing. This would not be the case in a reduced scale model tested at $1\,g$.

8.3.2.4 Inter-region transfer numbers.   Many natural soils exhibit heterogeneity which can be modelled by a two-region soil structure. When following this approach, the mechanisms governing transport in the soil are assumed to constitute the controlling processes within each region, together with inter-region transfer.

For problems concerning contaminant transport, it has been demonstrated that inter-region contaminant transfer gives rise to the phenomenon of physical, non-equilibrium sorption (Li *et al.*, 1994). This suggests that inter-region transfer is an important mechanism in effecting time-dependent transport behaviour in soils.

The process of inter-region contaminant transfer is governed by the velocity contrast between the two regions.

For a relatively high-velocity contrast, the fast-flowing region may be considered as a 'mobile' region, where contaminant transport is due to advection and dispersion, while the slow-flowing region may be considered as a stagnant or 'immobile' region, where contaminant transport is by diffusion alone. Under these conditions, contaminant transfer between the two regions is dominated by transverse, inter-region diffusion (Coats and Smith, 1964), and the controlling dimensionless group is:

$$I_1 = \frac{\alpha h^2}{D_T} \qquad (8.36)$$

where α is the inter-region contaminant mass transfer rate, h is the transverse dimension of a typical immobile region, and D_T is an effective, transverse diffusion coefficient.

For a low-velocity contrast, local velocity variations between the two regions dominate, and the controlling dimensionless group is given by

$$I_2 = \frac{\alpha D_h}{u^2} \tag{8.37}$$

The process of inter-region contaminant transport will be similar in model and prototype if $I_m = I_p$, in other words if $I_r = 1$. Because α describes a rate process, the scaling laws derived in section 8.3.1 indicate that α should scale according to $\alpha_r = 1/N^2 \, (\equiv 1/t_r)$. Following on from this, it is straightforward to demonstrate that the inter-region transport numbers, I_1 and I_2, will be equivalent in a centrifuge model and the prototype provided that $h_r = N$, and $D_{Tr} = D_{hr} = 1$. This suggests that the modelling of inter-region contaminant transfer may be possible during centrifuge testing; this point is discussed further in section 8.3.3.4.

8.3.2.5 Capillary effects number. The scaling laws derived in section 8.3.1 considered transport under conditions of saturated flow. However, in many cases, transport events take place in zones of unsaturated soil, where capillary effects are important.

The phenomenon of capillarity is described by the capillary effects number

$$C_a = \frac{\rho g h_c d}{\sigma} \tag{8.38}$$

where h_c is the capillary head and σ denotes surface (or interfacial) tension.

For the correct modelling of flow processes in the unsaturated zone, C_a must be identical in model and prototype ($C_{am} = C_{ap}$).

For a centrifuge model test conducted using prototype soil and fluids, $\rho_r = d_r = 1$. The capillary head, h_c, will scale as a macroscopic length, hence $h_{cr} = N$. Thus $c_{ar} = 1$, suggesting that similitude of capillary effects will be achieved between a centrifuge model and the prototype; this would not be the case in a reduced scale model tested at the Earth's gravity g.

8.3.3 Validity of scaling

In order that the same principles of conservation and transport govern both a centrifuge model and the prototype, certain conditions need to be observed. This section describes the circumstances under which these conditions will be observed, with specific reference to the work reviewed in this chapter.

8.3.3.1 Assumption of laminar flow. Throughout the previous discussion, it has been assumed that Darcy's law could be used to describe fluid motion in both the model and prototype.

Evidence suggests that Darcy's law is only valid where flow is laminar and viscous forces are predominant (Bear, 1972), in other words in the region where $R_e \leqslant 10$ (refer to section 8.3.2.1).

Since flow velocities under natural conditions are typically low, the condition $R_e \leqslant 10$ is likely to describe most prototype situations. However, because flow velocities are scaled in a centrifuge model, $R_{em} = N\,R_{ep}$. Thus, it is always necessary to confirm that $R_{em} \leqslant 10$, in order that fluid motion is suitably modelled during centrifuge testing.

8.3.3.2 Modelling contaminant dispersion.

The scaling factors presented in Table 8.1 assume that the process of contaminant dispersion will be similar in model and prototype (i.e. that $D_{dr}^* = D_r = 1$). Discussion in section 8.3.2.2 indicated that this would occur under two conditions. Firstly, when the model Peclet number remains below unity, i.e. when diffusion dominates the dispersion mechanism. Secondly, when the characteristic length effecting dispersion is scaled between model and prototype.

For homogeneous, porous media, the characteristic length effecting dispersion is of the order of the grain size of the medium (Bear, 1979). Scaling of this length requires that centrifuge model tests be performed using a material having an average grain size N times smaller than that of the prototype material. The scaling of grain sizes would be difficult to achieve in most cases, and, in addition, could affect the similitude of certain material parameters between model and prototype. Consequently, centrifuge models are most often constructed using prototype soil. As a result, when $P_{em} > 1$, the dispersion of contaminant within a centrifuge model is generally greater than that within the prototype. An investigation into the modelling of contaminant dispersion in homogeneous media is reviewed in section 8.4.4.

For large, non-uniform, field scale systems, l_c describes the spatial variability of soil properties in the average direction of flow. Thus, scaling of l_c in a centrifuge model involves reducing the dominant prototype heterogeneities by a factor N. Even assuming that we can accurately describe the nature of our prototype, the construction of a centrifuge model that faithfully incorporates reduced scale, prototype heterogeneities is likely to prove difficult. In addition, spatial variability in many field systems extends over distances that are not capable of being modelled using most centrifuge facilities. Thus, centrifuge model tests may not easily provide analogues for the phenomenon of field-scale dispersion.

8.3.3.3 Modelling internal solute sources and sinks.

In describing contaminant transport in section 8.2.3, it was presumed that the only contaminant–solid-phase interaction was linear equilibrium sorption. A necessary assumption of this model is that surface chemical reactions occur 'instantaneously', or at least 'rapidly'. In practice, a number of contaminant species are believed to obey rapid, linear equilibrium laws. However, caution is required when applying the adjective 'rapid' to both field and centrifuge time-scales. When modelling real transport problems, surface reactions that occur over periods of less than 24 hours may often be termed 'rapid' (Gerritse, personal communication, 1990). However, when viewed in terms of average centrifuge testing times ($t_m \leqslant 30$ hours), a chemical reaction that occurs over a period of 24 hours can no longer be considered 'rapid'.

At present, the nature of many microscopic, surface reactions remains largely unknown. For most adsorption processes, the actual reaction mechanisms must be known in order to determine the proper centrifuge scaling relationships. Clearly, further work is required in this area to identify the limits for centrifuge modelling of reactive species.

The derived scaling factors also require that linear decay processes (either within the fluid phase or upon solid surfaces) should occur N^2 times faster in a centrifuge model than in the prototype. If centrifuge model tests are carried out using the prototype contaminant, the relationship $\lambda_r = 1/N^2$ will generally not hold, because linear decay normally involves a 'real time' process. It is therefore improbable that prototype decay processes can be correctly scaled in a centrifuge model.

8.3.3.4   Modelling inter-region contaminant transport.   In section 8.3.2.4, we identified two mechanisms for controlling inter-region contaminant transport in heterogeneous soils, namely transverse, inter-region diffusion (characterised by the dimensionless number I_1) and flow variations (characterised by I_2).

Where the mechanism of inter-region diffusion dominates, it was noted that correct scaling of inter-region contaminant transport could be achieved if (i) the inter-region diffusion coefficient D_T was the same in model and prototype and (ii) the transverse dimension of a typical 'immobile' region, h, was scaled by a factor N.

Provided that the same materials and fluids are used in model and prototype, similitude of D_T is likely to be achieved. However, the feasibility of scaling h will depend upon the problem under investigation. If h is known, and of adequate dimension, the correct scaling of h may be possible. Otherwise, if true scaling of h is not possible, the effects of inter-region transfer in a centrifuge model will be less significant than those in the prototype.

Where local velocity variations dominate, it was noted that correct scaling of inter-region contaminant transport could be achieved if the value of D_h were the same in model and prototype. As previously discussed, this will arise when $P_e < 1$ in both model and prototype. Thus, in contrast to the above case, the correct scaling of inter-region transfer can be achieved using a model constructed with the prototype soil, provided that molecular diffusion dominates contaminant dispersion.

The potential for centrifuge modelling of contaminant transport with physical, non-equilibrium, sorption is coupled with the feasibility of scaling inter-region transport. An investigation into the potential for modelling physical, non-equilibrium sorption under low flow conditions is reviewed in section 8.4.5.

8.3.4   Benefits of centrifuge modelling

The main requirement for model testing in any area of engineering is to ensure that key dimensionless groups are equivalent in model and prototype. As indicated above, no physical modelling technique is likely to achieve correct scaling of *all* processes in the prototype. Instead, effort is usually directed towards obtaining similitude of those processes that determine the general behaviour of the prototype.

Centrifuge testing offers two major advantages in the physical modelling of geo-environmental problems, namely: (i) the technique provides a means of accelerating transport processes in soil (recall that $t_r = N^2$), thus enabling observation of phenomena that occur over long prototype times in short model test times; and (ii) the technique simulates *all* self-weight effects in the prototype.

Self-weight enters problems that involve subsurface transport in three respects. First, the movement of liquid pollutants through soil is heavily dependent upon material properties such as soil permeability and matrix porosity. The material properties of many compressible soils are a function of both the stress level and stress history of the soil. Thus, the achievement of identical stress at homologous points in model and prototype, is likely to lead to a true distribution of material properties throughout the soil model.

Secondly, physical transport due to gravitational forces will enter all transport problems concerning convective heat transfer and/or interaction between fluids of contrasting density.

Thirdly, geo-environmental problems frequently involve flow, at some stage, in groundwater systems near or below a water table, where zones of capillary rise and gradients of total potential are governed by gravity.

Thus, there are many conditions under which correct physical modelling of sub-surface transport will not be achieved unless self-weight is fully accounted for in the modelling technique; this fact was demonstrated to some extent during the preceding discussion on dimensionless numbers.

The previous section described several instances where the validity of the derived scaling laws may be violated. For some classes of problem, the scale effects introduced by failing to model certain processes, such as contaminant dispersion and linear decay, are likely to be small. Under these circumstances, observations made with regard to the centrifuge model can be considered representative of the prototype behaviour. Under other circumstances, centrifuge model tests can be regarded as independent transport events, producing data under repeatable and controlled laboratory conditions. Irrespective of the validity of the scaling laws, such data can still be used to test and verify existing conceptual models' transport codes.

8.3.5 Instrumentation

The centrifuge modelling of transport processes demands miniature instrumentation capable of measuring fluid pressure, soil temperature and contaminant concentration during centrifuge flight, as described below.

8.3.5.1 Pressure transducers.

In the majority of work reported below, fluid pressures in the centrifuge models and ancillary equipment were monitored by commercially available Druck PDCR81 miniature pore-water pressure transducers. The Druck PDCR81 has a head 12.5 mm long and 6.5 mm in diameter. A 2.2 mm diameter Teflon tube, extending from the back of the head, carries the transducer electrical lead to the electrical driver. A detailed description of the Druck PDCR81 and its operating system is given by, for example, Hensley (1989).

For some of the work reported below, the electrical system driving the Druck PDCR81 was adapted to allow the measurement of temperature by the transducer. This enabled transducer readings to be corrected for temperature rise, and allowed the simultaneous measurement of pressure and temperature at a singular point in the soil. The operating principle of the PDCR81 under these conditions is described by Savvidou (1984).

8.3.5.2 Thermocouples.

Where the simultaneous measurement of temperature and pressure was not required, temperature changes in the work reported below were monitored by Type-K (nickel–chrome) thermocouples. The thermocouples were manufactured by spot-welding nickel–chrome and nickel–aluminium PTFE-insulated wires. Protection of the weld was achieved by coating the joint in a thin cover of heat resistant epoxy resin. Cold junction references for the thermocouples were provided by commercially available electrical units.

8.3.5.3 Resistivity probes.

To date, much of the work involving the centrifuge modelling of contaminant transport, has concerned the migration of an ionic pollutant (sodium chloride) through saturated deposits of soil. In the majority of this work, the instrumentation used to monitor contaminant progress has constituted a resistivity (or conductivity) probe system.[4]

In most soil sediments, electrical conduction takes place through the pore fluid only, the soil grains themselves providing an insulated matrix to the flow of electricity. Thus, for a particular soil, variation in sediment resistivity is a

---

[4] The resistivity of a solution is the reciprocal of its electrical conductivity; thus the terms resistivity and conductivity describe an equivalent physical property.

function of soil porosity and the chemical composition of the pore fluid (Bates, 1991). The movement of ionic contamination (and in some cases organic contamination) can therefore often be mapped using measurements of electrical resistivity.

The measurement of electrical resistivity in soil may be performed using either a four-electrode or a two-electrode system. In a four-electrode system, an alternating current is passed through an outer pair of current electrodes and a potential difference is measured between an inner pair of potential electrodes. In a two-electrode system, the current and potential electrodes are identical.

Several styles of miniature resistivity probe have been developed for use in centrifuge modelling, together with several modes of electrical driver for probe energisation.

Figure 8.3(a) illustrates a miniature four-electrode resistivity probe, designed and developed at Cambridge University Engineering Department. The probe electrodes consist of four 0.5 mm thick high-purity silver washers, with external and internal diameters of 10 mm and 3 mm, respectively. The

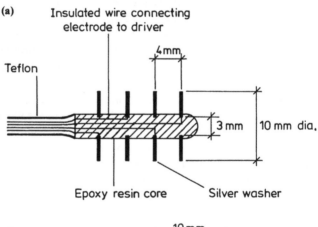

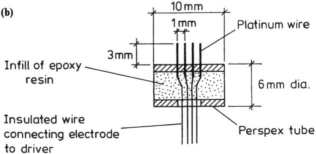

Figure 8.3 Miniature four-electrode resistivity probes. (a) Section through four-washer probe; (b) section through four-pin probe; (c) multiplex driver/receiver.

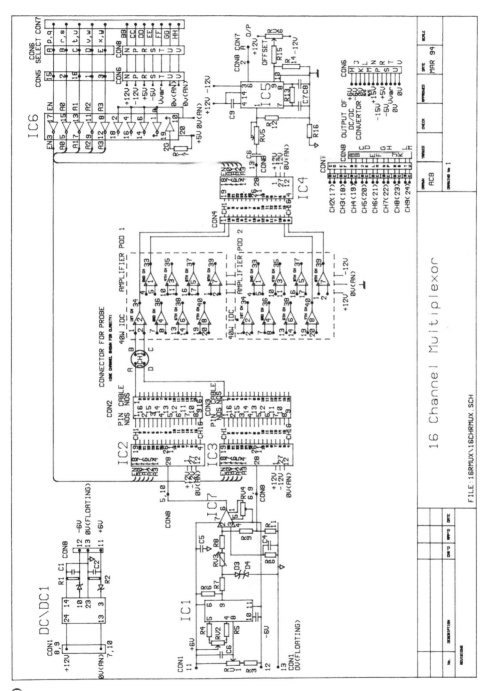

16 Channel Multiplexor

FILE: 16RMUX\16CHRMUX.SCH

electrodes are spaced at 4 mm intervals along an insulated core of epoxy resin. The overall length of each probe head is 16 mm. A 2.2 mm diameter Teflon tube, extending from the back of the probe head, carries wires from each electrode to the electrical driver.

Figure 8.3(b) illustrates an alternative four-electrode resistivity probe, also designed and developed at Cambridge University Engineering Department. The probe comprises four platinum wires spaced at 1 mm intervals along a 10 mm long, 6 mm diameter Perspex tube. The probe electrodes extend 3 mm from the outer surface of the tube. Four twisted PTFE-insulated wires, extending from the back of the probe head, connect the electrodes to the electrical driver.

An electrical driver for the miniature four-electrode probes is shown in Figure 8.3(c). The driver uses a multiplexing technique to individually energise and measure the signals from each probe, thus avoiding any interference that might occur between electrical fields of neighbouring probes.

Figure 8.4(a) shows a two-pin miniature combined resistivity and temperature probe, designed and developed at The University of Western Australia. The probe electrodes constitute two 0.5 mm diameter platinum–rubidium (Pt–Rb) wires. The electrodes are buried in a cylindrical core of porous ceramic material, having an average grain size of 0.5 mm. Temperatures at the electrodes are monitored by a nickel–chrome thermocouple, bonded to the base of the core. The head of the probe has an average diameter of 5 mm, and an overall length of 6 mm. Each probe electrode is connected to the driver by a fine PTFE-insulated wire.

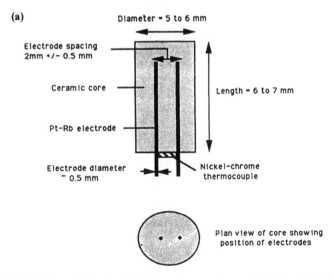

Figure 8.4 Miniature two-electrode resistivity and temperature probe: (a) section through probe (b) multiplex driver/receiver. (From Hensley and Savvidou, 1993.)

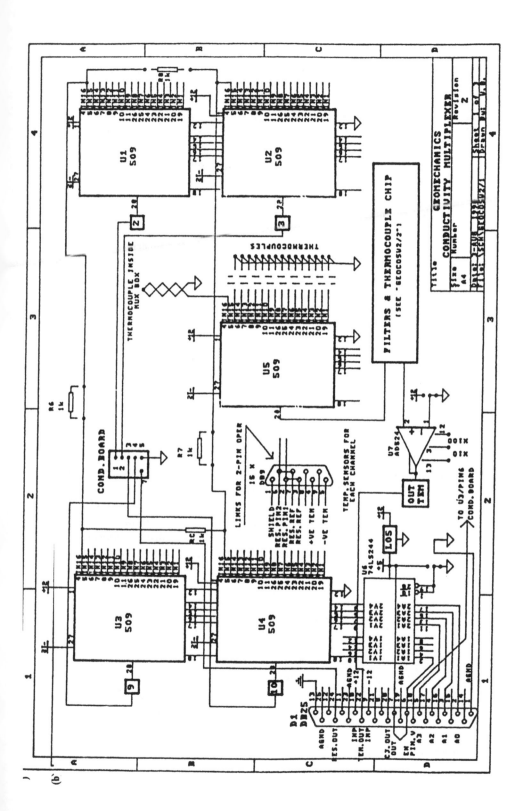

Because the medium surrounding the electrodes of the miniature two-pin probe remains invariant, probe readings change only with the chemical composition of fluid passing through the probe head. Accordingly, probe calibrations do not change with soil type or soil porosity.

Figure 8.4(b) presents an electrical driver for the two-pin combined resistivity and temperature probe. Again, the driver uses a multiplexing technique to individually energise and interrogate each probe.

8.4 Modelling contaminant transport

This section reviews work from seven different projects concerning centrifuge modelling of contaminant transport. The work covered by the projects encompasses problems involving non-reactive, contaminant transport in zones of uniform or heterogeneous soil, including problems involving unsaturated and immiscible fluid flow.

8.4.1 Migration from a landfill site

This work investigated non-reactive, contaminant migration from a landfill sited in a saturated, soil deposit of finite depth. The work was conducted on the balanced arm centrifuge at the University of Cambridge, UK, and is reported by Hensley (1988, 1989) and Hensley and Schofield (1991).

8.4.1.1 Model tests. The prototype problem chosen for the study is shown in Figure 8.5. Both depletion of contaminant in the landfill with time, and a constant landfill concentration, were studied during the test series.

A 0.6 M sodium chloride solution was used to represent the model pollutant. The principal geological deposit was formed from a uniform, saturated silt, having an average volume porosity of 0.38 and a hydraulic conductivity of 8.8×10^{-8} m/s. The model aquifer was constructed from a graded, satur-

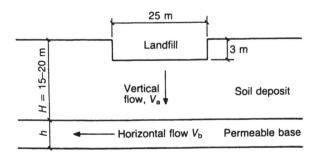

Figure 8.5 Section through prototype of problem. (From Hensley and Schofield, 1991. © Institution of Civil Engineers.)

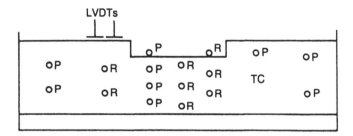

Figure 8.6 Typically instrumented centrifuge model: P, pore-pressure transducer; R, resistivity probe; TC, thermocouple; LVDT, linear variable differential transformer. (From Hensley and Schofield, 1991. © Institution of Civil Engineers.)

ated sand, having an average volume porosity of 0.34 and a hydraulic conductivity of 7.3×10^{-4} m/s.

A schematic representation of a typically instrumented centrifuge model is given in Figure 8.6: miniature resistivity probes of the type shown in Figure 8.3(a) were used throughout the test series.

Figure 8.7 illustrates the service arrangements for a typical centrifuge model. The advective seepage velocity through the silt layer, and the horizontal seepage velocity in the base aquifer, were controlled by water levels within standpipes connected to the model. Alteration of standpipe overflow heights allowed for the variation of model seepage velocities between tests.

Each centrifuge test followed the same general procedure. Initially the centrifuge was started, and the speed increased in stages to $100\,g$. The flow pattern within the model was then established using a 'clean' (uncontaminated) fluid supply to the landfill. After approximately 2 hours, when transducer readings indicated steady-state conditions in the model, the uncontaminated fluid was removed from the landfill, and replaced by a sodium chloride solution while the centrifuge was in flight. For the remaining duration of the test, the landfill was supplied with either a standard sodium

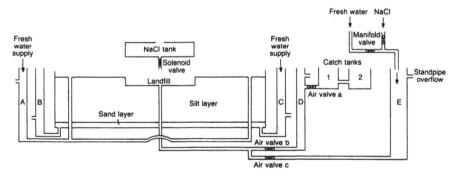

Figure 8.7 Typical package service arrangements. (From Hensley and Schofield, 1991. © Institution of Civil Engineers.)

chloride solution, representing a constant concentration of pollutant in the landfill, or uncontaminated fluid, representing a depletion of landfill concentration with time.

A total of seven centrifuge tests were undertaken during the study. The tests encompassed a number of groundwater seepage velocities and a variety of landfill boundary conditions. Results from the tests were compared with predictions from several theoretical analyses.

8.4.1.2 Comparison of test data and theoretical predictions. The following gives a comparison of data from the fifth centrifuge test in the series, Test PJH5, and predictions by a commercially available computer transport code, POLLUTE (Rowe and Booker, 1985).

Test PJH5 modelled a landfill sited in a 14.8 m deep silt deposit, underlain by a sandy aquifer of 2.5 m depth. A relative vertical specific discharge of 0.965 m/year was set through the medium beneath the landfill site, and a relative horizontal specific discharge of 230.21 m/year was imposed along the base aquifer.

Experimental data and theoretical predictions are compared in Figure 8.8. The results are presented in prototype time ($t_p = N^2 t_m$, where $N = 100$) and normalised with respect to c_s, the volume-based concentration of the model pollutant.

Figure 8.8(a) presents a record of contaminant concentration in the landfill site. For an initial 17.5 years, the contaminant held in the landfill was maintained at a maximum concentration, c_s. Thereafter, the concentration of landfill contaminant was allowed to diminish with time.

Figures 8.8(b) and (c) compare experimental and theoretical data at measurement points located directly below the landfill site. Clearly, excellent agreement was obtained between observed and predicted data at all points, including that point located within the base aquifer.

8.4.1.3 Summary. This work was undertaken to investigate the technique of centrifuge modelling of contaminant transport processes. The tests performed were the first of this nature to investigate a realistic problem. Prototype times of up to 30 years were modelled during the test series. Long-term *in situ* tests of this nature would have been costly and extremely expensive to perform, and may have provided little direct control over boundary conditions.

The centrifuge test results were compared with data from commercially available transport codes. These comparisons served to illustrate the enormous potential of the centrifuge in providing good quality experimental data for the verification of mathematical transport models.

8.4.2 Remediation of contaminated land

This section discusses the modelling of site remediation using flow reversal

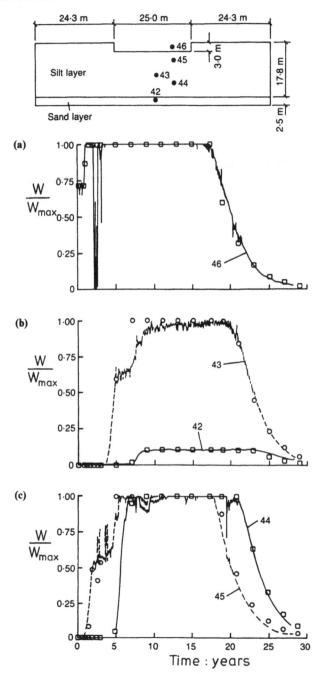

Figure 8.8 Predicted concentration changes during centrifuge test PJH5 using POLLUTE; theoretical data are represented by discrete points: (a) at probe 46 (landfill site); (b) at probes 42 (sand layer) and 43 (silt layer); (c) at probes 44 and 45 (silt layer); inset, transducer locations and prototype dimensions. (From Hensley and Schofield, 1991. © Institution of Civil Engineers.)

techniques. The work was conducted using the balanced arm centrifuge at the University of Cambridge, UK, and is reported by Hellawell *et al.* (1993a,b) and Hellawell (1994).

8.4.2.1 Centrifuge model and preparation. A possible technique for re-mediation of contaminated land in relatively permeable soils is 'pump and treat'. This process involves flushing the site with clean water to dilute the contaminant, followed by an appropriate surface treatment.

The centrifuge test described here modelled the hydrodynamic clean-up of a two-dimensional site using flow reversal. The prototype is shown in Figure 8.9. The remediation involved pumping clean water via two wells through a layer of uniformly contaminated silt. The effects of flow reversal were also investigated during the test.

The soil used was reconstituted fine silica flour. A 0.1 M sodium chloride solution was used to represent a non-reactive pollutant. An initially uniformly contaminated site was created by reconstituting the silica flour with the 0.1 M solution. When consolidated, the silt had a volume porosity of 0.45 and a hydraulic conductivity of 3.0×10^{-7} m/s.

The service arrangements for the test are shown in Figure 8.10. Two wells ($20 \times 152 \times 180$ mm) were constructed from 2.5 mm grade G Vyon. These were placed 93 mm above the base of the liner at a separation of 280 mm. Fluid was supplied to the wells via a feed pipe from two standpipes.

Contaminant concentrations were monitored using two forms of resistivity probes; those of the types shown in Figures 8.3(b) and 8.4(a), respectively.

8.4.2.2 Test procedure. Initially, all air valves and clean water feeds were closed. The centrifuge was started and the speed increased in stages to the target acceleration of 100 g. Following sample consolidation, the wells were

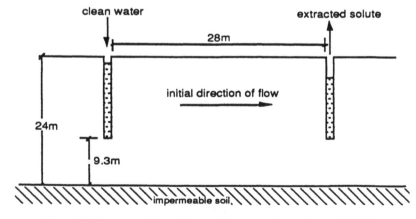

Figure 8.9 Prototype of centrifuge model. (From Hellawell *et al.*, 1993b.)

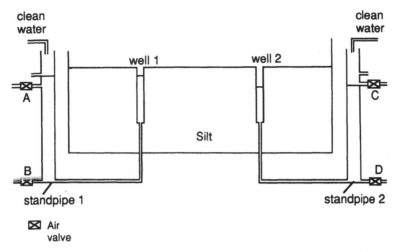

Figure 8.10 Service arrangements for the centrifuge test. (From Hellawell *et al.*, 1993b.)

flushed with clean water, and clean-up of the soil commenced with fluid being drawn from well 1 to well 2.

After sufficient cleaning, the flow was reversed, and fluid was drawn from well 2 to well 1.

8.4.2.3 Test results. The results of the centrifuge test were compared with theoretical predictions from a model developed by Hellawell *et al.* (1993b).

Figure 8.11 gives a comparison of theoretical predictions and centrifuge test data. For the purpose of analysis, it was assumed that the area between the wells had a uniform hydraulic gradient, and could be approximated using a one-dimensional model. During the experiment, the flow direction was reversed once the clean front had reached the centre of the model.

The data presented in Figure 8.11 demonstrate reasonable agreement between the experiment and analysis.

8.4.2.4 Summary. The work presented here modelled the clean-up of a large-scale prototype over five years. Centrifuge test results provided physical data that compared favourably with theoretical predictions, thereby demonstrating the feasibility of using a geotechnical centrifuge to investigate aspects of site remediation.

8.4.3 Density-driven flow and hydrodynamic clean-up

This work concerned the study of a contaminant transport problem involving density-driven flow and hydrodynamic clean-up. The work was conducted using the balanced arm centrifuge at the University of Cambridge, UK, and is reported by Hellawell and Savvidou (1994) and Hellawell (1994).

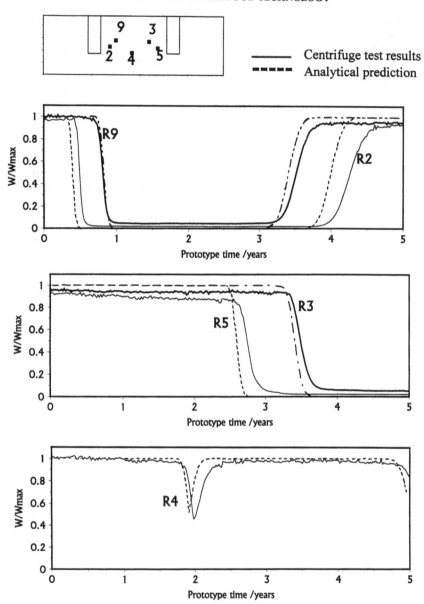

Figure 8.11 Comparison of theoretical predictions and results from the centrifuge test. (From Hellawell *et al.*, 1993b.)

8.4.3.1 Centrifuge model. A test was performed to model the self-weight migration and clean-up of a dense pollutant. The prototype considered was a 10 m wide landfill leaking dense pollutant through its base into a homogeneous soil layer (Figure 8.12). The level of fluid in the landfill was constant,

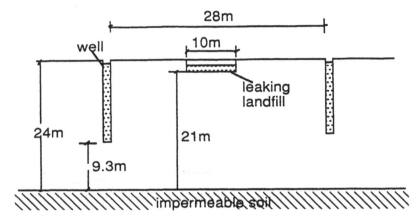

Figure 8.12 Prototype for centrifuge model. (From Hellawell and Savvidou, 1994.)

at the same height as the water table in the surrounding soil. Two sampling wells were positioned 28 m apart on either side of the landfill.

The soil deposit was modelled by a uniform saturated silt. A 0.1 M sodium chloride solution was used as the model pollutant. Contaminant concentrations during testing were monitored with resistivity probes of the type shown in Figure 8.3(b). Thermocouples were attached to the probes to record variations in soil temperature.

The wells were constructed from moulded Vyon filled with coarse sand. During the test, they were supplied with fresh water from two standpipes connected to the package. Figure 8.13 illustrates the service arrangements for the test. A wave gauge was used to determine the fluid depth in the landfill. A control loop from the gauge signal operated a peristaltic pump and solenoid valve, which regulated the contaminant supply from an overhead reservoir.

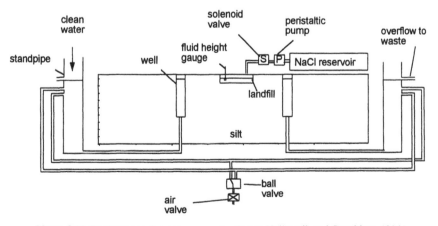

Figure 8.13 Package service arrangements. (From Hellawell and Savvidou, 1994.)

When active, the control loop automatically maintained a constant head of fluid in the landfill.

8.4.3.2   Test procedure.   Initially, the air valve was closed and a small amount of contaminant was placed in the landfill. The centrifuge was then started and the speed increased in stages to the target acceleration of 100 *g*. Upon reaching the target acceleration, the control loop operating the feed of pollutant to the landfill was activated, and the landfill was filled with contaminant. The sample was then allowed to consolidate under self-weight. After approximately 15 minutes, pore pressures in the sample reached equilibrium.

For the following 26.5 hours, the plume from the landfill was allowed to develop, after which the air valve was opened and the cleaning phase began. During clean-up, fluid was drawn from one well to another. The introduction of flow reversal was achieved through switching the ball valve.

8.4.3.3   Comparison of test data and theoretical predictions.   The results of the centrifuge test were compared with predictions using the finite difference transport code HST3D (Heat and Solute Transport in Three Dimensions; Kipp, 1987).

Figure 8.14 gives a comparison of theoretical predictions and centrifuge test data. The results are presented in prototype time. All stages during the test were considered during numerical modelling.

The effect of density driven migration can be seen at probes 3 and 6. These probes were positioned at approximately the same depth. However, density effects, which caused greater contaminant movement at the edges of the landfill, resulted in the latter probe detecting the plume first.

The movement of the plume during clean-up is demonstrated by the effects of flushing on probes 7, 3 and 5. At the end of the test, probe 3 was beginning to detect the clean front, whereas high contaminant concentrations were still being detected at probe 5. Thus, after 5.7 years of pumping, only half the site had been cleaned.

In general, reasonable agreement was obtained between measured values and theoretical data during the phases of both plume creation and clean-up. The slight discrepancy in the slopes of the curves was attributed to small numerical dispersion.

8.4.3.4   Summary.   The geotechnical centrifuge was used for modelling the migration and clean-up of a dense pollutant originating from a landfill site. True representation of all density-related effects was achieved during modelling.

The centrifuge test results demonstrated that hydrodynamic clean-up is not satisfactory in less than highly permeable soils, and validated the HST3D

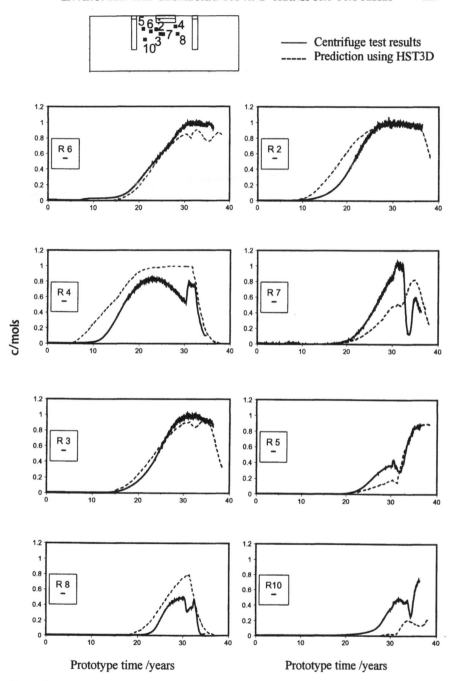

Figure 8.14 Comparison of theoretical predictions using HST3D and results from the centrifuge test. *Note:* pumping phase commenced at prototype time of 30.5 years. (From Hellawell and Savvidou, 1994.)

transport code for the migration and clean-up processes investigated under the conditions of the test.

8.4.4   Contaminant dispersion in uniform soil

This work investigated the potential of centrifuge testing for modelling dispersive processes in uniform, saturated soil. The work was conducted using the centrifuge facility at The University of Western Australia, and is reported by Hensley and Randolph (1994).

8.4.4.1   Background.   In uniform, saturated soil, the best representation of the dispersion mechanism has proved to be the macroscopic equation of dispersion. For one-dimensional flow, this equation is given by (Fried, 1975):

$$\frac{\partial c}{\partial t} = D_h \frac{\partial^2 c}{(\partial z')^2} \tag{8.39}$$

where z' gives the position of an observer moving with the average flow velocity u ($z' = z - ut$). Note that the equation is written for a non-reactive contaminant species.

For continuous injection of a contaminant species with concentration c_0 at location $z = 0$, the solution to equation (8.39) can be approximated by a normal distribution function with mean $m = ut$, and standard deviation $s = (2D_h t)^{0.5}$. Thus, the magnitude of D_h can be determined from a profile of pollutant concentration at time t, provided that the appropriate initial and boundary conditions have been observed.

Under conditions of one-dimensional flow, it is common to describe D_h in the form (Freeze and Cherry, 1979):

$$D_h = T'D_d + a_L u \tag{8.40}$$

where T' is the tortuosity of the porous medium and a_L is a longitudinal dispersion coefficient that quantifies the medium's heterogeneity (non-uniformity).

The first term on the left-hand side of equation (8.40) characterises that contribution made to contaminant dispersion by molecular diffusion, whereas the second term characterises that contribution made by mechanical dispersion.

8.4.4.2   Experimental programme.   To explore whether centrifuge testing could be used to obtain dispersion coefficients in soils, a series of three centrifuge tests were undertaken using a coarse sand having an average particle diameter of 0.85 mm.

The arrangement of the centrifuge model is shown in Figure 8.15. A saturated sand layer with an average volume porosity of 0.36 and a height of 158 mm was constructed over a 53 mm thick silt layer underlain by a

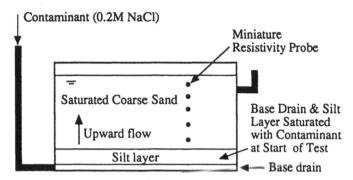

Figure 8.15 Form of centrifuge model. (From Hensley and Randolph, 1994.)

permeable base drain. The upward hydraulic gradient through the sample was controlled by water levels within standpipes connected to the model. A fluid head difference of 7.5 mm was maintained across the model during all tests. Fluid velocities were varied between each test, by adjusting the centrifugal force (g level) acting on the model.

All tests modelled the continuous injection of a non-reactive pollutant species into an initially uncontaminated soil. At the start of each test, the base drain and silt layer were saturated with an 0.2 M sodium chloride solution, while the sand layer remained saturated with fresh water. Upward transport of the contaminant through the initially 'clean' sand layer was continuously monitored by resistivity probes of the type shown in Figure 8.4(a). The coefficient of hydrodynamic dispersion for chloride through the sand was determined from an observed profile of pollutant concentration at a fixed time, t. The results of the centrifuge test series are given in Table 8.2.

Figure 8.16 presents a comparison of dispersion coefficients measured using standard laboratory techniques (Campbell, 1991) with those measured during the centrifuge tests. Good agreement was observed between the coefficients, suggesting that centrifuge modelling is a viable means of obtaining dispersion coefficients through soils.

8.4.4.3 Predicting dispersion in a prototype. Theoretically, the pollutant concentration profiles observed during a centrifuge test may be transformed

Table 8.2 Summary of centrifuge tests

Test	g level	Velocity, u (m/s)	D_h(m^2/s)
1	150 g	9.66×10^{-6}	1.84×10^{-8}
2	53 g	3.55×10^{-6}	0.99×10^{-8}
3	100 g	6.33×10^{-6}	1.48×10^{-8}
Prototype	1 g	0.06×10^{-6}	0.11×10^{-8}

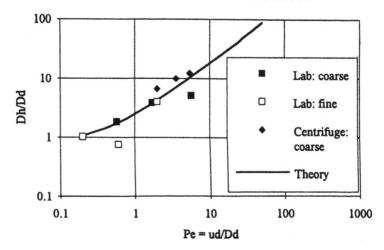

Figure 8.16 Comparison of dispersion coefficients; theory given by a relationship proposed by Perkins and Johnston (1963). (From Hensley and Randolph, 1994.)

into predicted prototype concentration profiles by means of the scaling laws presented in Table 8.1. Figure 8.17 presents normalised prototype concentration profiles at $t = 2.4$ years predicted using data from the centrifuge test series, together with the prototype concentration profile, obtained from a solution to the macroscopic equation of dispersion (equation 8.39).

Figure 8.17 shows agreement between all profiles at the 'mean' concentration $c/c_0 = 0.5$. However, there is obvious disagreement between the predicted 'spread' of each profile about the mean.

This disagreement can be explained by the variance in the dispersion

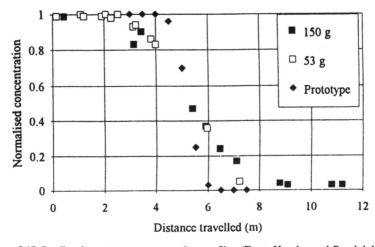

Figure 8.17 Predicted prototype concentration profiles. (From Hensley and Randolph, 1994.)

coefficients presented in Table 8.2. The scaling laws given in Table 8.1 require similitude of dispersion between model and prototype (see section 8.3.3.2). Clearly, such similitude was not observed, and we note that the higher the dispersion coefficient characterising the centrifuge model, the wider the 'spread' of the predicted profile, and the greater the disparity between the prediction and the actual prototype.

8.4.4.4 A scaling law for pollutant spread.   The data above confirm that there is a scaling error in modelling dispersion when $D_{hm} \neq D_{hp}$. This section discusses whether the error can be accounted for by means of a separate scaling factor for pollutant spread.

Following the procedure given in section 8.3.1, the macroscopic equation of dispersion (equation 8.39) can be written for the prototype as:

$$\frac{c_r}{t_r}\frac{\partial c_m}{\partial t_m} = D_{h_r}D_{h_m}\frac{c_r}{(z_r')^2}\frac{\partial^2 c_m}{\partial (z_m')^2} \tag{8.41}$$

and for the model as:

$$\frac{\partial c_m}{\partial t_m} = D_{h_m}\frac{\partial^2 c_m}{\partial (z_m')^2} \tag{8.42}$$

By comparing equation (8.41) with equation (8.42), we find that for correct scaling:

$$\frac{1}{t_r} \equiv \frac{D_{h_r}}{(z_r')^2} \tag{8.43}$$

It was established in section 8.3.1 that $t_r = N^2$. Hence:

$$\frac{D_{h_r}}{(z_r')^2} = \frac{1}{N^2} \tag{8.44}$$

The coordinate z' characterises the spread of a pollutant profile about the mean concentration $c/c_o = 0.5$. Thus, if we know D_{h_r}, we can establish the appropriate scaling ratio for transcribing the pollutant spread observed in a centrifuge model to the prototype situation.

For centrifuge model tests carried out using the same soil as the prototype, equation (8.40) can be used to estimate that[5]:

$$D_{h_r} = \frac{D_{hp}}{D_{hm}} = \frac{T'D_d + a_L u_p}{T'D_d + a_L(Nu_p)} \tag{8.45}$$

Thus, for tests conducted within the region where $T'D_d \gg a_L(Nu_p)$, molecular diffusion will dominate ($P_{em} < 1$), D_{h_r} will equal unity and z_r' will equal N. When this arises there will be no scaling error in modelling dispersion, and the

---

[5] If the centrifuge model is constructed using prototype soil, it is reasonable to assume that the values of T' and a_L will be identical in model and prototype.

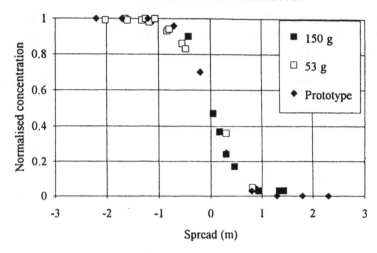

Figure 8.18 Corrected concentration profiles. (From Hensley and Randolph, 1994.)

scaling factor for the spread of pollutant (z'_r) will be identical to that for scaling length (z_r). However, for tests conducted outside this region $(P_{em} > 1)$, the value of z'_r will lie between N and $N^{1/2}$, and the scaling factor for pollutant spread will be less than that for length.

The data presented in Figure 8.17 were plotted assuming, incorrectly, that the same centrifuge scaling factor applies to pollutant spread and length. Figure 8.18 presents the same centrifuge test data 'corrected' to account for the proper scaling of pollutant spread. This figure demonstrates the validity of the scaling law derived for pollutant spread, and suggests that, even if a scaling error is introduced when modelling dispersion on a centrifuge, we can still relate model concentration profiles to a prototype.

8.4.4.5 Summary. This work established the centrifuge as a new experimental tool for modelling dispersive processes, and demonstrated that centrifuge test data can be used to predict prototype behaviour, even when similitude of dispersion processes is not achieved between model and prototype.

8.4.5 *Physical, non-equilibrium sorption in heterogeneous soil*

This work investigated the potential for centrifuge modelling of contaminant transport with non-equilibrium, physical sorption. The work was conducted using the centrifuge facility at The University of Western Australia, and is reported in Li (1993) and Li *et al.* (1993, 1994a,b).

8.4.5.1 Background. Physical, non-equilibrium sorption arises when 'apparent' time-dependent sorption is caused by physical factors, such as heterogeneous hydraulic conductivity.

The transport of a non-reactive contaminant undergoing physical non-equilibrium sorption is mathematically described by a two-region soil model (Coats and Smith, 1964). Although numerous research workers have shown that a two-region model can adequately describe transport with physical, non-equilibrium sorption (see for example De Smedt and Wierenga, 1979), the model's application is, in practice, restricted by the difficulties in determining the model parameters. In addition, it has been shown that a two-region model is mathematically identical to the 'two-site sorption model', used to describe chemical, non-equilibrium sorption (Parker and van Genuchten, 1984). Consequently, the mechanism of non-equilibrium sorption (i.e. whether it is due to chemical or physical phenomena) cannot be determined by the models themselves through curve fitting. It therefore seems likely that physical modelling of this class of problem may be of considerable practical use.

8.4.5.2 Model tests. The feasibility for centrifuge modelling of physical non-equilibrium sorption was investigated during six tests, that incorporated the principle of 'modelling of models'.

The model tests were all performed using an artificially constructed, two-region soil structure. Polyethylene porous cylinders (PPC; diameter 13.5 mm, length 10 mm) were used to simulate one of the soil regions. The cylinders were interspersed uniformly in a homogeneous silt at a volume percentage of 14%. The consolidated silt had a hydraulic conductivity of 3.5×10^{-6} m/s and volume porosity of 0.43; the polyethylene cylinders had a hydraulic conductivity of 4.2×10^{-5} m/s and a volume porosity of 0.5. Sodium chloride was used as the non-reactive contaminant.

Two models were used during the test series, Model 1 and Model 2. Both models were constructed in a plastic lined aluminium column of 140 mm diameter. For Model 1, the soil column was 240 mm long, which was twice that used for Model 2. Model 1 was run under three different centrifuge accelerations ($g_1 = 10\,g$, $g_2 = 30\,g$, $g_3 = 50\,g$), while Model 2 was run under three corresponding accelerations ($g'_1 = 20\,g$, $g'_2 = 60\,g$, $g'_3 = 100\,g$). The configuration of each model is shown in Figure 8.19. Hydraulic heads at the surface and base were maintained by two pre-set overflows. The concentration of the effluent discharged at the base was measured using a resistivity probe of the type shown in Figure 8.4(a).

After set-up, the model was placed in a centrifuge strong box and accelerated for 60 minutes to allow consolidation of the soil. Once sufficient consolidation time had elapsed, a pulse of contaminant was introduced to the surface of the soil column. Following on from this, and for the remaining duration of the test, deionised water was continuously fed to the soil surface. At the end of the test, the breakthrough curve (BTC) of chemical transport

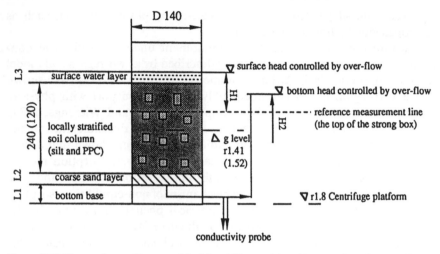

Figure 8.19 Form of centrifuge model. $L1 = 120$ mm, $L2 = 45$ mm, $L3 = 13$ mm (6.5 mm). $H1 = 43$ mm (-77 mm), $H2 = 7$ mm (-95 mm). $\triangle h = 36$ mm (18 mm). The values in parentheses are for Model 2. Negative values are below the reference measurement line. (After Li *et al.*, 1994a.)

through the soil column was determined from data logged from the probe at the column base.

During all tests, a steady, but low, flow rate was maintained through the soil column. A maximum Peclet number of 0.8 was imposed during the test series, suggesting that the inter-region transfer giving rise to physical non-equilibrium sorption was dominated by local velocity variations.

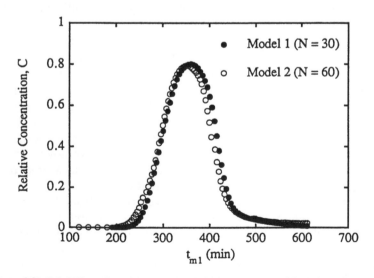

Figure 8.20 'Modelling of models' tests: Set 2 (MT3 and MT4). (After Li *et al.*, 1994a.)

8.4.5.3   Test results.   The two centrifuge models designed for the purpose of 'modelling of models' were separated by a scale factor of 2. Thus, the data from Model 2 could be plotted together with the data from Model 1, by scaling the time of Model 2's BTC according to $t_{m1} = t_{m2} \times 2^2$.

Figure 8.20 shows an example of a scaled BTC curve from Model 2, plotted together with the relevant BTC data of Model 1. The results clearly demonstrate the feasibility of modelling Model 1 by Model 2. This fact was observed for all comparisons.

To estimate the error of the physical modelling approach, the first three temporal moments of the BTCs for both models were calculated.[6] The calculations showed that these moments for the two models were close enough to each other, to suggest that the transport processes in both models were similar.

8.4.5.4   Summary.   The work presented here demonstrated that centrifuge modelling of physical, non-equilibrium sorption in a two-region soil structure under low-flow conditions is possible. This suggests that centrifuge testing may provide a useful technique for estimating and predicting non-equilibrium contaminant transport in soils. Further, by comparing non-reactive and reactive chemical transport experiments, the technique may also offer a means of separating and quantifying the physical and chemical causes of non-equilibrium.

8.4.6   Flow and contaminant transport in the unsaturated zone

This section presents work carried out to evaluate centrifuge modelling of non-reactive contaminant transport in partially saturated soil. The work was conducted using the centrifuge facility at Queen's University, Ontario, Canada, and is reported in Cooke (1991, 1993) and Cook and Mitchell (1991a,b).

8.4.6.1   Background.   For unsaturated soils, both the hydrodynamic dispersion coefficient, D_h, and the hydraulic conductivity, K, vary hysteretically with the volumetric fluid content, θ (Bear, 1979). Thus, the mathematical problems involved in modelling unsaturated contaminant transport are considerable, and, in some cases, unsurmountable. Thus, it seems likely that physical modelling may provide a useful alternative to predicting contaminant transport in the unsaturated zone.

8.4.6.2   Model tests.   An evaluation of the potential for centrifuge modelling of unsaturated contaminant transport was conducted using a 100 mm

---

[6] These three moments physically represent the mass conservation, the mean movement and the variance, respectively.

diameter, soil column container, that had been fabricated to allow examination of each 20 mm section following testing. Water tanks at the top
and base of the column apparatus were designed to maintain pre-set fluid
levels, while effecting in-flight, flow measurement. A sodium chloride solution
was used as a non-reactive contaminant. All model tests were conducted at
Peclet numbers of less than unity.

Soil columns were prepared by compacting a homogeneous, silty sand in
layers, followed by saturation from the base under a vacuum-induced hydraulic gradient. Two heights of soil columns were prepared: for 50 g models a
300 mm column was prepared, while for 100 g models a 160 mm column was
prepared. On a prototype scale, the columns represented a homogeneous, *in
situ* soil profile nominally 15 m thick.

Before the start of each test, a soil column was placed in the centrifuge
strong box, brought up to the target acceleration (50 or 100 g), consolidated,
and then subjected to a permeability test under a hydraulic gradient of unity.
The bottom water level, representing the groundwater table, was maintained
constant at a level 1–2 cm above the column base for both the permeability
test and the drainage test described below.

Following the permeability test, the centrifuge was stopped, the upper
water supply tank disconnected, and both tanks emptied. The sample was
then accelerated again, and allowed to drain for the equivalent of 15 000
hours prototype time (1.5 hours at 100 g or 6 hours at 50 g). At this stage,

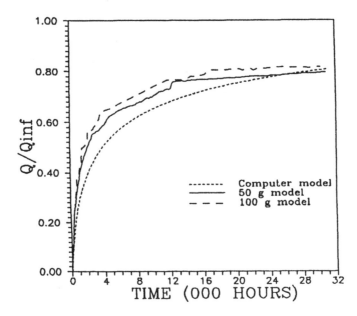

Figure 8.21 Relative outflow rates of 50 g, 100 g and computer models. (From Cooke and
Mitchell, 1991b.)

either 4.6 cm^3 (for the 100 g model) or 9.2 cm^3 (for the 50 g model) of 10% sodium chloride solution was injected into the centre of the soil surface, using an in-flight dispensing apparatus.

The sample was then centrifuged for the equivalent of an additional 15 000 prototype hours before the test was stopped.

At the end of each test, the soil column was partitioned into 20 mm slices, and each slice was analysed for soil moisture content and contaminant concentration.

8.4.6.3 Verification of results. The centrifuge test results were verified by comparison with a computer model based on the Richards equation (Richards, 1931), and by comparison of different scale models at the prototype scale (i.e. by 'modelling of models').

A comparison of results for the soil drainage process is given at the prototype scale in Figures 8.21 and 8.22.

Typical results from the column drainage test (Figure 8.21) show good agreement between the computer model and both the 50 g and 100 g centrifuge models. The moisture content profiles at an intermediate stage of drainage (Figure 8.22) show comparably good agreement between the centrifuge models, with a slightly poorer fit with the computer model. Overall, the results are good enough to suggest that the soil drainage process was faithfully modelled during centrifuge testing.

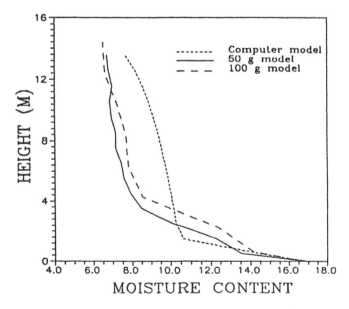

Figure 8.22 Moisture content profiles for 50 g, 100 g and computer models. (From Cooke and Mitchell, 1991b.)

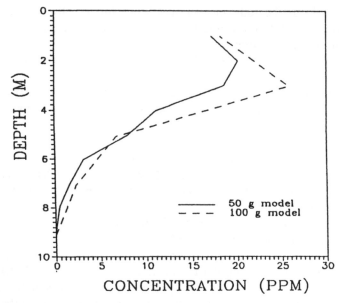

Figure 8.23 Tracer concentrations for 50 g and 100 g model profiles. (From Cooke and Mitchell, 1991b.)

Contaminant concentration profiles are compared at the prototype scale in Figure 8.23. Again, close agreement is observed between the centrifuge models. This gives confidence in the feasibility of modelling unsaturated contaminant transport using a geotechnical centrifuge.

8.4.6.4 Summary. Centrifuge modelling in partially saturated soil has verified that groundwater flow and moisture suction phenomena can be modelled during centrifuge testing. In addition, data obtained on the migration of a non-reactive contaminant have indicated that centrifuge modelling may provide a useful means of studying transport processes in the unsaturated zone.

8.4.7   Multi-phase flow

This work investigated centrifuge modelling of immiscible fluid transport in soil. The work was conducted using the centrifuge facility at the University of Colorado, Boulder, USA and is reported by Illangasekare *et al.* (1991).

8.4.7.1 Background. In addition to miscible fluids, hazardous wastes are also frequently found in the form of non-aqueous phase liquids (NAPLs), which are sparingly soluble in water. In recent years, there has been much interest in predicting how these chemicals move through unsaturated zones of an aquifer, towards the water table.

If a sharp penetration front is maintained, NAPL infiltration through unsaturated soil is described by an equation analogous to the Richards equation for water infiltration. Thus, the capillary effects and scaling laws that govern unsaturated flow are equally applicable to this class of problem.

8.4.7.2 Model tests.   Model tests were conducted to investigate the movement of a NAPL through unsaturated soil following a surface spill. Soil samples were prepared in a transparent Plexiglass cylinder, having an inside diameter of 215 mm and a height of 310 mm. A porous stone was placed at the base of the cylinder so that air could escape freely during infiltration. Graduations at one-inch intervals were marked on the cylinder to map the position of the infiltrating front with time.

Two soil types were used in the experiments, namely a uniform, fine sand packed to a volume porosity of 0.44, and a uniform silt having a volume porosity of 0.52 and a 9.5% degree of water saturation.

The organic chemical Soltrol 220 was used as the lighter than water non-aqueous phase fluid (LNAPL). In order to visually observe its migration during testing, the Soltrol was marked with a red dye.

In the case of the sand, spill simulations were conducted for two types of initial conditions, namely (1) completely dry sand and (2) sand at a uniform residual saturation of 2%.

At the start of each test, the organic fluid was applied at the top of the soil column to an initial depth of 89 mm. The front locations were then recorded as a function of time under a gravity of 20 g.

In all tests, a reasonably sharp front was maintained between the advancing front and the displaced fluids.

8.4.7.3 Test results.   The results from the centrifuge tests were compared with results from prototype testing (Figure 8.24), and with theoretical predictions from a simple three parameter model developed by Doshi (1989) (Figure 8.25).

Figure 8.24 demonstrates good agreement between the scaled centrifuge test data and the prototype data. This indicates that centrifuge testing could provide a viable means of modelling certain multi-phase flow processes.

Good agreement is also observed between the centrifuge test data and predictions from the three-parameter model (Figure 8.25). This gives further confidence in the modelling technique, and establishes that the three-parameter model can adequately describe NAPL infiltration under the conditions of the test series.

8.4.7.4 Summary.   Results from this work demonstrated the feasibility of using centrifuge modelling techniques for studying immiscible fluid flow through soil. In addition, test data were successfully used to verify a simple multiphase flow model for sharp infiltration.

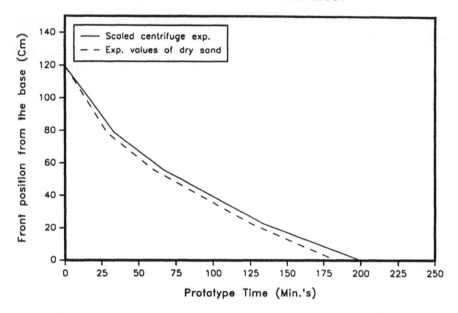

Figure 8.24 Comparison of prototype results and converted results from centrifuge experiment to 1 *g*. (From Illangasekare *et al.*, 1991.)

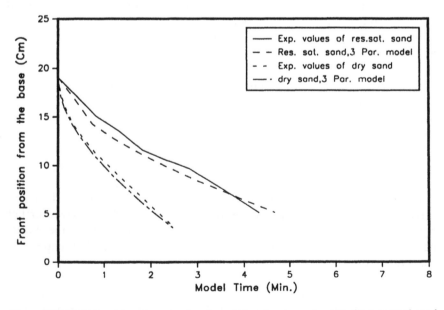

Figure 8.25 Fluid front location as a function of time for dry and residually saturated sand. (From Illangasekare *et al.*, 1991.)

8.5 Modelling heat and mass transport

This section reviews work from three different projects that concerned the centrifuge modelling of heat and mass transport. The work covered by the projects includes problems involving both conductive and convective heat transport, together with problems involving combined heat and pollutant transport.

8.5.1 Heat transfer and consolidation

This work investigated heat transfer and consolidation in a fine-grained soil. The work was conducted using the balanced arm centrifuge at the University of Cambridge, UK, and is reported by Savvidou (1984), Maddocks and Savvidou (1984) and Britto et al. (1989).

8.5.1.1 Background. Low-permeability, fine-grained soils tend to be purely conducting media, where heat transfer takes place through the 'static' pore fluid and the solid particles themselves.

Temperature rises in a purely conducting medium cause both the solid particles and the pore fluid to expand. In a fully saturated soil, the pore fluid expands more than the solid particles, thereby generating positive excess pore pressures. Counteracting this is the dissipation of such pressures, as fluid is squeezed out of the permeable medium during consolidation. If the net increase in excess pore pressure is sufficient, effective stresses in the soil can be reduced to levels at which cracks and fissures form. The prospect of this may be of concern in problems relating to, for example, the disposal of heat generating wastes in low-permeability environments.

A coupled analysis of heat conduction and consolidation in compresssible media was developed by Booker and Savvidou (1985). The centrifuge model tests described here were performed to study temperatures and pore pressure changes in a fine grained soil surrounding a heat source, and to verify that analysis.

8.5.1.2 Model tests. The prototype for the centrifuge model tests concerned a canister 0.6 m in diameter and 6 m long, containing a source of heat-generating waste. It was envisaged that the hot canister would be installed in a submerged bed of saturated clay, which was to form a long-term barrier against waste migration to the environment.

A series of five tests was carried out to investigate temperature and pore-pressure rises around model heated canisters installed in submerged beds of kaolin clay. The clay was contained in an 850 mm diameter, 400 mm high steel tub shown in Figure 8.26. In each test, two canisters were used. The model canisters were installed in the clay during centrifuge flight using a pulley

(a)

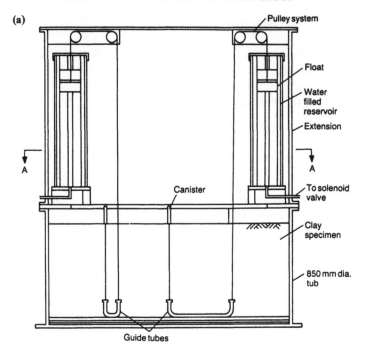

(b)

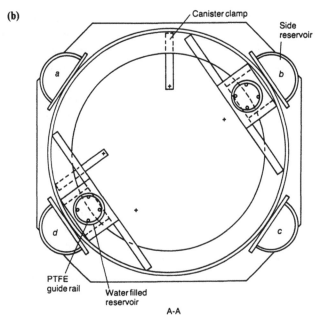

A-A

Figure 8.26 Apparatus developed to install model canisters in clay during centrifuge tests: (a) vertical section; (b) cross-section A–A. (From Britto *et al.*, 1989. © Institution of Civil Engineers.)

system. Following installation, a heating element housed in the body of the canister was used to raise its surface temperature to a maximum of 90°C.

Temperatures and pore pressures following installation and heating were monitored during each test by thermocouples bonded to the surface of each canister, and by Druck PDCR81 pressure transducers, adapted to measure both temperature and pressure at points in the soil.

All tests were conducted at a centrifuge acceleration of $100\,g$. Thermal Rayleigh numbers during the test series remained below $4\pi^2$, signifying that heat transfer in all models was by pure conduction alone.

8.5.1.3 Test results.   Here, results are presented from two centrifuge tests representative of the series, Test CS5 and Test CS4. The results are compared with data from a finite element program, HOT CRISP (Britto, 1984), that incorporates the analysis of Booker and Savvidou (1985).

Test CS5 concerned the installation and subsequent heating of a canister in a bed of normally consolidated clay. Results from the test are compared with theoretical predictions from HOT CRISP in Figure 8.27.

The comparisons indicate that the temperature distribution around the heater was well predicted by the analysis. In addition, it is clear that the analysis provided a reasonable estimate of the pore-pressure rises generated in the normally consolidated clay during heating.

Test CS4 modelled the installation and subsequent heating of a canister in a bed of lightly overconsolidated clay with an OCR of 1.5. A comparison of test data with predictions from HOT CRISP is given in Figure 8.28.

Once again, the comparisons show that the temperature field around the canister was well predicted by the analysis. However, in contrast, it is clear that pore-pressure rises generated in the lightly overconsolidated clay during heating were under-predicted by the analysis.

Site investigation following the end of Test CS4 revealed the presence of horizontal cracking centred mainly at the top and bottom ends of the canister. The analysis, which did not take into account remoulding of the clay around the canister during installation, did not predict this effect.

8.5.1.4 Summary.   Data from the test series were used to verify an analysis based on conductive heat transfer and consolidation. Comparisons of experiment and analysis demonstrated that the analysis could reasonably predict the behaviour observed in a normally consolidated clay. However, this was not the case in a lightly overconsolidated clay.

Horizontal cracking in the soil surrounding a cylindrical heat source was observed at the end of one centrifuge test. This effect was not predicted by the analysis. Physical modelling on the centrifuge thereby provided invaluable information on phenomena related to stress history and stress levels which was not provided by theoretical modelling techniques.

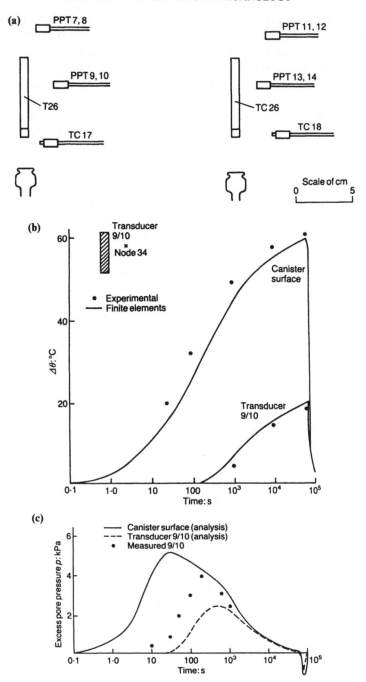

Figure 8.27 Results from centrifuge model test CS5: (a) transducer positions; (b) temperature variation with time; (c) comparison of predicted and observed pore pressures. (From Britto *et al.*, 1989. © Institution of Civil Engineers.)

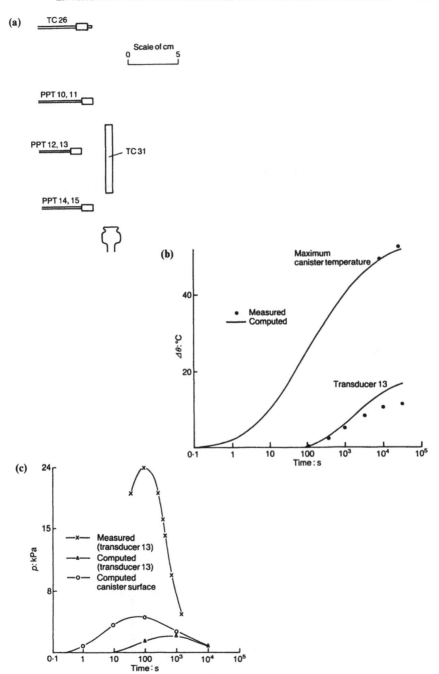

Figure 8.28 Results from centrifuge model test CS4: (a) transducer positions; (b) temperature variation with time; (c) comparison of predicted and observed pore pressures. (From Britto *et al.*, 1989. © Institution of Civil Engineers.)

8.5.2 Convective heat transfer

The work presented here assessed the potential of centrifuge modelling for studying convective heat transfer in coarse grained soil. The work was conducted using the Rowe geotechnical centrifuge at the University of Manchester, UK, and is reported by Savvidou (1988).

8.5.2.1 Background.

At thermal Rayleigh numbers greater than approximately 40 ($R_{aT} > 4\pi^2$), heat-transfer problems become unstable, and free convection of the pore fluid arises (Elder, 1976). In such cases, the transference of heat through the soil is dominated by pore fluid movement and not by conduction.

This study was undertaken to demonstrate the use of the centrifuge in modelling convective heat transfer, and to verify the scaling laws related to this class of problem.

8.5.2.2 Physical modelling.

Experiments were performed on four models to investigate transport mechanisms around a heat source buried in coarse-grained soil.

The soil used for the study was a saturated, uniform sand with a grading fraction size from 1.18 to 2.36 mm. The model sand layer was prepared at a void ratio of 0.5, corresponding to a hydraulic conductivity of 5×10^{-4} m/s.

During testing, the sand was contained in a steel box with dimensions $560 \times 560 \times 460$ mm high. The box was insulated externally by fibre glasss layers, and internally by 30 mm thick Perspex sheets. A long cylindrical heater

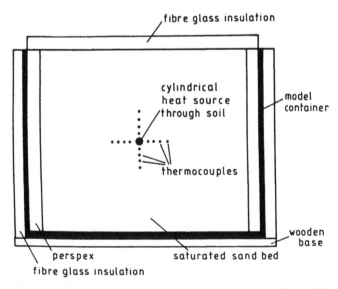

Figure 8.29 Cross-section of model container. (From Savvidou, 1988.)

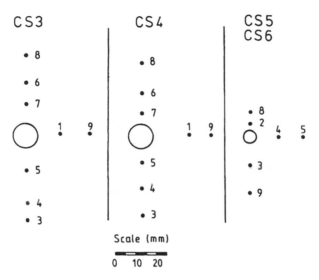

Figure 8.30 Thermocouple locations around heat source. (From Savvidou, 1988.)

was positioned horizontally and centrally in the box, and thermocouples were mounted at set distances from the heater surface (Figure 8.29). The positions of the thermocouples during the different tests are shown in Figure 8.30.

Two different diameter heaters were used during the test series, namely a 12.7 mm diameter heater and a 6.35 mm diameter heater. In all centrifuge tests, a constant power of 700 W was supplied to the heat source, which raised the heater surface temperature by about 40°C. It was found that a similar surface temperature change was achieved at 1 g when the power supplied to the heater was about 175 W.

Models were tested on the centrifuge under an acceleration of either 100 or 50 g. For comparison, 1 g tests were also performed on each model. Each test lasted for a duration of approximately 3 hours, during which time temperatures on and around the heat source were monitored continuously by the thermocouples.

8.5.2.3 Results and discussion. In order to compare the different modes of heat transfer invoked during centrifuge tests and 1 g tests, results are presented from Model CS6. Model CS6 incorporated a 6.35 mm diameter heater and was tested under both 1 g and 100 g. At 1 g, the thermal Rayleigh number was given by $R_{aT} = 4.8$, whereas at 100 g $R_{aT} = 480$.

Figure 8.31(a,b) shows the temperature changes occurring at the early stages of heating in the model when it was tested under 1 g and 100 g, respectively. In the 100 g test, the thermocouples positioned above the heat source registered sudden temperature increases, whereas the remaining

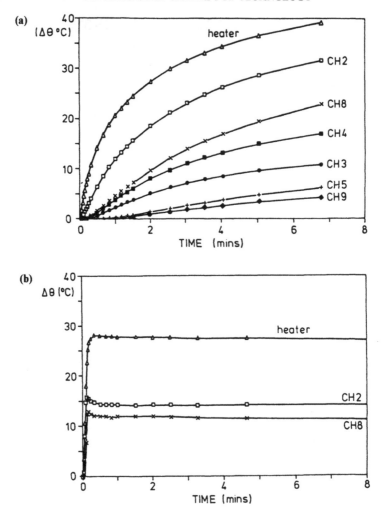

Figure 8.31 Variation of temperature with time: (a) Model CS6, 1 *g*; (b) Model CS6, 100 *g*. (From Savvidou, 1988.)

thermocouples recorded temperature changes between 0° and −1°C; these are not shown in the figure. Conversely, in the 1 *g* test, a more gradual temperature change was detected by all thermocouples surrounding the heat source during the same time period.

The experimental results were found to be consistent with a form of buoyancy induced, convective heat transfer at 100 *g*, and with predominately conductive heat transfer at 1 *g*. These observations verify that different heat transfer mechanisms are invoked at different thermal Rayleigh numbers, thereby confirming that R_{aT} must be identical in model and prototype if similitude of the heat transfer mechanism is to be achieved.

For the purpose of verifying scaling laws, 'modelling of models' was conducted. Results from one model, Model CS4, which was tested under $50\,g$ with a 12.7 mm diameter heater, were compared with results from Model CS6, which had been tested under $100\,g$ with a 6.35 mm diameter heater (Figure 8.32).

The experimental data were used to calculate the time taken for the heat front to advance vertically upwards in both models. Comparison of the computed times confirmed that $t_r = N^2$.

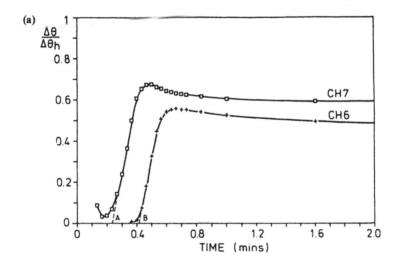

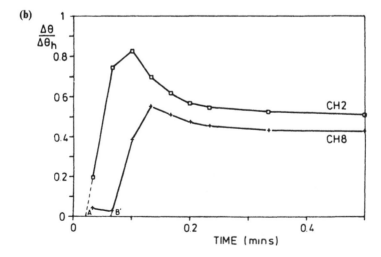

Figure 8.32 Normalised temperature variation above the heat source: (a) Test CS4, $50\,g$; (b) Test CS6, $100\,g$. (From Savvidou, 1988.)

8.5.2.4   Summary. The results of this study established the value of centrifuge modelling for problems involving convective heat transport. Comparison between models tested under $100\,g$ and under $1\,g$ demonstrated a
distinct difference in the mechanism by which heat transfer took place in the
two cases. Furthermore, an attempt to verify scaling laws by the 'modelling of
models' was successful, confirming that $t_r = N^2$.

8.5.3   Heat and pollutant transport

This work was undertaken to investigate coupled heat and contaminant
transport in soil surrounding a hot, buried waste source. The work was
conducted using the centrifuge facility at The University of Western
Australia, and is reported by Hensley and Savvidou (1992, 1993).

8.5.3.1   Background. During coupled heat and contaminant transport, the
behaviour of many prototype problems is governed by hydraulic instability.
As discussed in section 8.3.2.3, the onset of instability affecting this class of
problem is related to the magnitude of the effective Rayleigh number, R_e.
   During centrifuge testing, the behaviour of a model under different effective
Rayleigh numbers can be observed simply by varying the g level under which
the model is tested. Thus, the centrifuge offers an ideal tool for investigating
the mechanisms of combined heat and contaminant transport under different
conditions.

8.5.3.2   Physical modelling. Experiments were performed in two models to
investigate effects around a combined heat and contaminant source buried in
coarse grained soil. The two models were constructed from uniform, saturated sand. The first model comprised a 'fine' sand, having an average particle

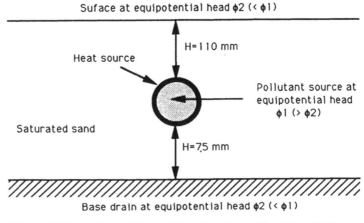

Figure 8.33 Form for model tests. (From Hensley and Savvidou, 1993.)

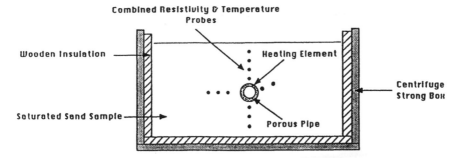

Figure 8.34 Schematic diagram of centrifuge model. (From Hensley and Savvidou, 1993.)

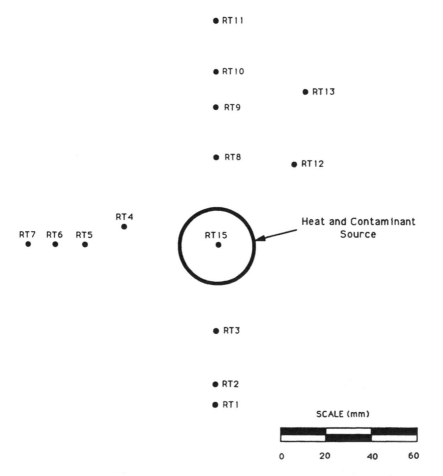

Figure 8.35 Location of combined resistivity and temperature probes. (From Hensley and Savvidou, 1993.)

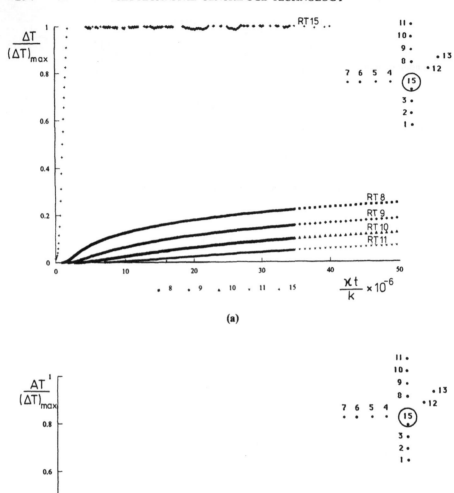

Figure 8.36 Results from centrifuge model test HST1B: (a) temperature rises within and above the source; (b) temperature rises below the source; (c) concentration change. (From Hensley and Savvidou, 1993.)

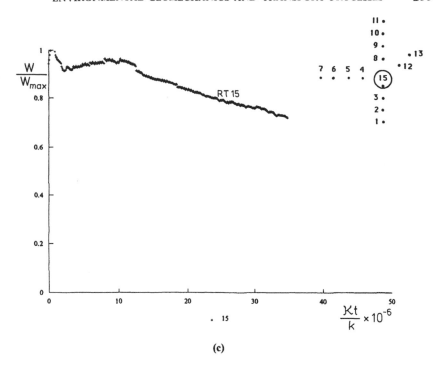

(c)

size of 0.1×10^{-3} m. The second model comprised a coarser grade sand, having an average particle size of 0.9×10^{-3} m.

The model form that was used for the study is shown in Figure 8.33. A sodium chloride solution, prepared at an average contaminant mass fraction $w = 10 \times 10^{-3}$ was used as a non-reactive pollutant.

The centrifuge models were enclosed in a rectangular strong box lined with a thermal insulator (Figure 8.34). A thin-walled horizontal pipe, constructed from sintered stainless steel, was used for the combined heat and contaminant source. The contaminant was introduced into each model during centrifuge flight along the sintered pipe. To raise the temperature of contaminant leaving the source, a heating element, constructed from two strands of twisted, insulated, solid copper wire, was wrapped around the perimeter of the pipe. Temperature rises at the crown and invert of the pipe were measured using thermocouples. Temperature and concentration rises in the pipe and surrounding soil were monitored by miniature combined resistivity and temperature probes of the type shown in Figure 8.4(a). Figure 8.35 presents a plan of the transducer locations.

Each centrifuge test followed the same general procedure. Initially the centrifuge was started and the speed increased in stages to the required acceleration. Equilibrium conditions within the centrifuge model were then established using a fresh water supply to the pipe. When all transducers

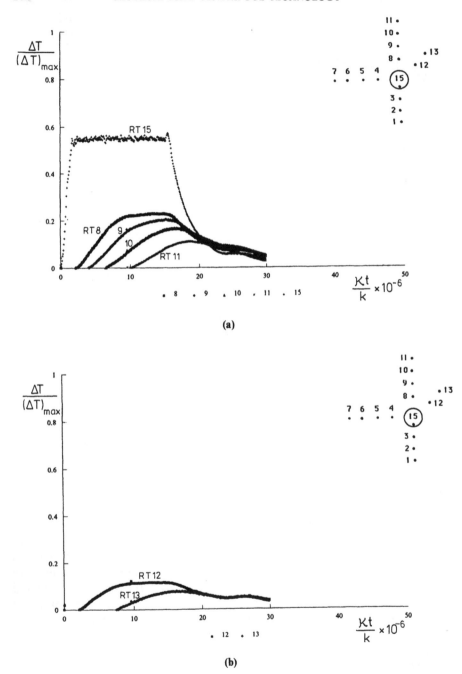

Figure 8.37 Results from centrifuge model test HST1C: (a) temperature changes within and above the source; (b) temperature changes diagonally above the source; (c) concentration changes. (From Hensley and Savvidou, 1993.)

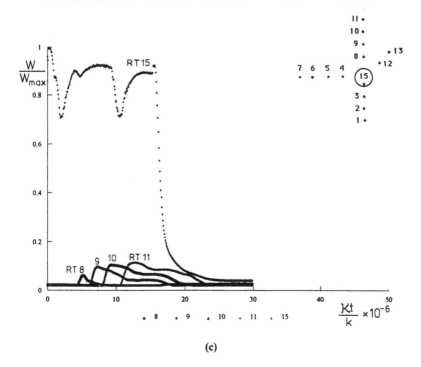

(c)

indicated steady-state conditions, the fresh water was removed from the pipe and replaced by a sodium chloride solution while the centrifuge was in-flight. Immediately after the pipe had been flushed with pollutant the heater was turned on. For the remaining duration of the test, the fluid temperature within the pipe was maintained at approximately 55°C. Signals from all instrumentation were monitored throughout each test. Each test was terminated when pollutant reached the boundaries of the model.

For comparison, 1 g tests were also performed on each model. The procedure followed during these tests was similar to that followed during a typical centrifuge test.

8.5.3.3   Test results.   In all, a total of six centrifuge tests and two 1 g tests were undertaken during the study. In order to illustrate the transport mechanisms observed under different conditions, results are presented from three tests representative of the series, namely tests HST1B, HST1C and HST2B.

Test HST1B was conducted in the fine sand at a low effective Rayleigh number ($R_{aT} \approx 0.5$ and $R_{as} \approx 75$). Typical normalised temperature changes recorded during the test are shown in Figure 8.36(a,b), while normalised concentration changes are presented in Figure 8.36(c). Throughout the test, the only recorded concentration changes occurred within the pipe itself.

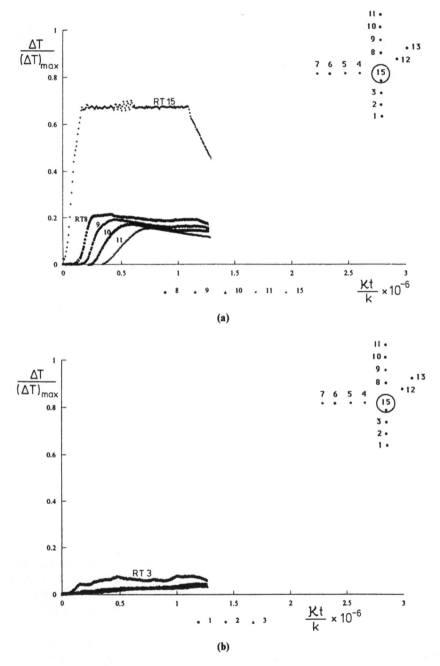

Figure 8.38 Results from centrifuge model test HST2B: (a) temperature changes within and above the source; (b) temperature changes below the source; (c) concentration changes within and above the source; (d) concentration changes below the source. (From Hensley and Savvidou, 1993.)

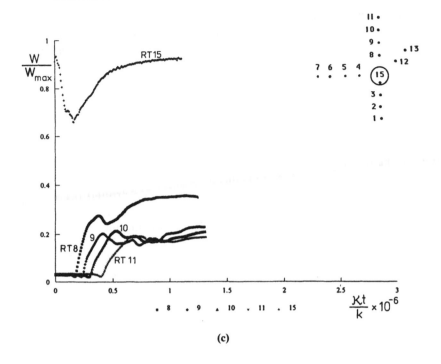

(c)

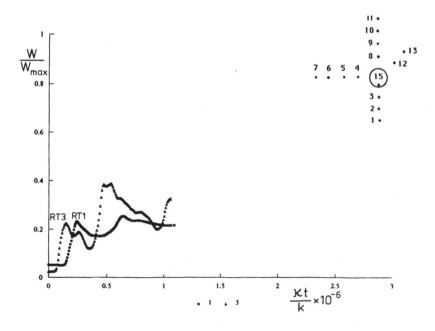

(d)

Both the temperature and concentration data gathered during test HST1B, indicate little, or no, movement of pore fluid around the source during testing. This suggests that hydraulic instability was not generated in the model under a low effective Rayleigh number.

Test HST1C was conducted in the fine sand at an intermediate effective Rayleigh number ($R_{aT} \approx 110$ and $R_{as} \approx 15\,500$). Normalised temperature changes recorded during the test are shown in Figure 8.37(a,b), while normalised concentration changes are given in Figure 8.37(c). During the test, changes in concentration were noted both within the pipe itself, and at probes buried in the soil above the pipe.

The nature of the temperature and concentration data gathered during test HST1C imply the instigation of free cellular convection around the source during testing. This suggests that some form of oscillatory instability (refer to Figure 8.2b) was generated in the model under an intermediate effective Rayleigh number.

Test HST2B was performed using the fine sand at a high effective Rayleigh number ($R_{aT} \approx 750$ and $R_{as} \approx 110\,000$). Temperature changes recorded during the test are shown in Figure 8.38(a,b), whereas recorded concentration changes are given in Figure 8.38(c,d).

The nature of the temperature data observed during test HST2B, suggest the formation of a slender plume above the source during the early stages of testing. This observation seems consistent with the nature of the concentration data recorded above the pipe. Conversely, the concentration data recorded below the pipe suggest some initial self-weight migration of pollutant, before any temperature change took place in the soil body. Overall, the results suggest that certain modes of direct instability (refer to Figure 8.2a) were generated in the model under a high effective Rayleigh number.

8.5.3.4 Summary. This work demonstrated that a geotechnical centrifuge is a useful tool in which to study problems of coupled heat and contaminant transport. Data from the test series confirmed distinct variations in the transport mechanisms under different effective Rayleigh numbers, and, additionally, identified certain boundaries of behaviour for a typical prototype problem.

8.6 Conclusions

This chapter has discussed centrifuge modelling of problems in environmental geomechanics, with particular reference to the study of sub-surface transport processes.

The fundamental mechanisms of sub-surface transport have been described, together with the principles and scaling laws related to the centrifuge modelling of these mechanisms. Following on from this, a review of centrifuge modelling work carried out within the discipline of environmental geomechanics has been presented. We have demonstrated the use of the centrifuge as a technique for modelling fluid flow, heat transport and contaminant transport in porous media and have shown that centrifuge test results can be used to calibrate numerical codes, study fundamental processes and even to predict prototype behaviour.

The wide spectrum of work carried out in this area has done much to establish the geotechnical centrifuge as a worthwhile tool for the physical modelling of environmental engineering problems. Centrifuge projects in this field are thereby only likely to increase as engineers come to realise the enormous contribution that centrifuge modelling can make towards the development of sound environmental management strategies.

References

Arulanandan, K., Thompson, P.Y., Kutter, B.L., Meegods, N.J., Muraleetharan, K.K. and Yogachandran, C. (1988) Centrifuge modelling of transport processes for pollutants in soils. *J. Geotech. Eng., ASCE*, **114** (2), 185–205.

Bates, C.R. (1991) Environmental geophysics in the US—a brief overview. *Geoscientist*, **2** (2), 8–11.

Bear, J. (1972) *Dynamics of Fluids in Porous Media*. American Elsevier, New York.

Bear, J. (1979) *Hydraulics of Groundwater*. McGraw-Hill, New York.

Bear, J. and Verruijt, A. (1987) *Modelling Groundwater Flow and Pollution*. D. Reidel, Dordrecht.

Bejan, A. (1984) *Convection Heat Transfer*. John Wiley, New York.

Booker, J.R. and Savvidou, C. (1985) Consolidation around a point heat source. *Int. J. Numer. Anal. Methods Geomech.*, **9**, 173–184.

Britto, A.M. (1984) *Finite Element Analyses of a Hot Canister Buried in Kaolin Clay*. Cambridge University Engineering Department Technical Report, CUED/D-Soils/TR191.

Britto, A.M., Savvidou, C., Maddocks, D.V., Gunn, M.J. and Booker, J.R. (1989) Numerical and centrifuge modelling of coupled heat flow and consolidation around hot cylinders buried in clay. *Géotechnique*, **38** (1), 13–25.

Campbell, R. (1991) A Study of Dispersion Processes in Porous Media. Project Report, Bachelor of Engineering, The University of Western Australia.

Celorie, J.A., Vinson, T.S., Woods, S.L. and Istok, J.D. (1989) Modelling solute transport by centrifugation. *ASCE J. Env. Eng.*, **115** (3), 73–84.

Coats, K.H. and Smith, B.D. (1964) Dead-end pore volume and dispersion in porous media. *Soc. Petrol. Eng. J.*, **4** (3), 73–84.

Cooke, B. (1991) Centrifuge Modelling of Flow and Contaminant Transport Through Partially Saturated Soils. PhD thesis, Queen's University, Kingston, Ontario, Canada.

Cooke, B. (1993) Physical modelling of contaminant transport in the unsaturated zone. *Geotechnical Management of Waste and Contamination* (eds R. Fell, T. Phillips and C. Gerrard), pp. 343–348. Balkema, Rotterdam.

Cooke, B. Mitchell, R.J. (1991a) Physical modelling of dissolved contaminant transport in an unsaturated sand. *Canad. Geotech. J.*, **28** (6), 829–833.

Cooke, B. and Mitchell, R.J. (1991b) Evaluation of contaminant transport in partially saturated soil. *Centrifuge '91* (eds H.Y. Ko and F.G. McLean), pp. 503–508, Balkema, Rotterdam.

Doshi, D. (1989) Modelling Vertical Migration of Nonaqueous Phase Liquid Waste in Unsaturated Soils. Dissertation. Louisiana State University, Baton Rouge, LO.

Elder, J.W. (1967) Transient convection in a porous medium. *J. Fluid Mech.*, **27**, 609–623.

Elder, J.W. (1976) *The Bowels of the Earth*. Oxford University Press, Oxford.

Freeze, R.A. and Cherry, J.A. (1979) *Groundwater*. Prentice-Hall, New Jersey.

Fried, J.J. (1975) *Groundwater Pollution*. Elsevier, Amsterdam.

262   GEOTECHNICAL CENTRIFUGE TECHNOLOGY

Hellawell, E.E. (1994) Modelling Transport Processes in Soil due to Hydraulic Density and Electrical Gradients. PhD Thesis, Cambridge University.

Hellawell, E.E. and Savvidou, C. (1994) A study of contaminant transport involving density driven flow and hydrodynamic clean up. *Proc. Int. Conf. Centrifuge '94*, pp. 357–362. Balkema, Rotterdam.

Hellawell, E.E., Savvidou, C. and Booker, J.R. (1993a) Clean up operations in contaminated land. *Proc. Int. Conf. Environmental Management, Geo-water and Engineering Aspects, Wollongong, Australia 8–11 February 1993*. Balkema, Rotterdam.

Hellawell, E.E., Savvidou, C. and Booker, J.R. (1993b) *Modelling of Contaminated Land Reclamation*. Cambridge University Engineering Department Technical Report, CUED/D-Soils/TR263.

Hensley, P.J. (1988) Geotechnical centrifuge modelling of hazardous waste migration. In *Land Disposal of Hazardous Waste: Engineering and Environmental Issues* (eds J.R. Gronow, A.N. Schofield and R.K. Jain), pp. 139–151. Ellis Horwood, Chichester.

Hensley, P.J. (1989) Accelerated Physical Modelling of Transport Processes in Soil. PhD Thesis, University of Cambridge.

Hensley, P.J. and Randolph, M.F. (1994) Modelling contaminant dispersion in saturated sand. *Proc. XIII Int. Conf. Soil Mech. Found. Eng., New Delhi*, pp. 1557–1560. Oxford and IBH Publishing Co. pvt Ltd, New Delhi, India.

Hensley, P.J. and Savvidou, C.S. (1992) Modelling pollutant transport in soils. *Austral. Geomech.*, **22** (July), 7–16.

Hensley, P.J. and Savvidou, C.S. (1993) Modelling coupled heat and contaminant transport in groundwater. *Int. J. Numer. Anal. Methods Geomech.*, **17**, 493–527.

Hensley, P.J. and Schofield, A.N. (1991) Accelerated physical modelling of hazardous waste transport. *Géotechnique*, **41** (3), 447–466.

Illangasekare, T.H., Znidarčić, Al-Sheridda, M. and Reible, D.D. (1991) Multiphase flow in porous media. *Centrifuge '91* (eds H.Y. Ko and F.G. McLean), pp. 517–523, Balkema, Rotterdam.

Kipp, K.L. (1987) HST3D: *A Computer Code for Simulation of Heat and Solute Transport in Three-dimensional Groundwater Flow Systems*. US Geological Survey Water-Resources Investigations Report 86-4095.

Lapwood, E.R. (1948) Convection of a fluid in a porous medium. *Proc. Camb. Phil. Soc.*, **44**, 508–521.

Li, L. (1993) Physical Modelling of Nonreactive Chemical Transport in Locally Stratified Soils. MEngSci Dissertation, The University of Western Australia, Nedlands.

Li, L., Barry, D.A., Hensley, P.J. and Bajracharya, K. (1993) Nonreactive chemical transport in structures soil: The potential for centrifuge modelling. In *Geotechnical Management of Waste and Contamination* (eds R. Fell, T. Phillips and C. Gerrard), pp. 425–431, Balkema, Rotterdam.

Li, L., Barry, D.A., Culligan-Hensley, P.J. and Bajracharya, K. (1994a) Mass transfer in soils with local stratification of hydraulic conductivity. *Water Resources Research* (in press).

Li, L., Barry, D.A. and Stone, K.J.L. (1994b) Centrifugal modelling of nonsorbing, non-equilibrium solute transport in locally inhomogeneous soil. *Canad. Geotech. J.* (in press).

Maddocks, D.V. and Savvidou, C. (1984) The effects of heat transfer from a hot penetrometer installed in the ocean bed. *Proc. Symp. Application of Centrifuge Modelling to Geotechnical Design, Manchester* (ed. W.H. Craig), pp. 336–355. Balkema, Rotterdam.

Neild, D.A. (1968). Onset of thermohaline convection in a porous medium. *Water Res. Res.*, **4** (3), 553–560.

Parker, J.C. and van Genuchten, M.Th. (1984) Determining transport processes from field tracer experiments. Virginia Agriculture Experiment Station, Bulletin 84-3. Virginia Polytechnic and State University, Blacksburg, Virginia.

Perkins, T.R. and Johnston, D.C. (1963) A review of diffusion and dispersion in porous media. *Pet. Trans. AMIE*, **228**, SPJ20, 70–84.

Richards, L.A. (1931) Capillary conduction of liquids through porous mediums. *Physics*, **1**, 318–333.

Rowe, R.K. and Booker, J.R. (1985) 1-D pollutant migration in soils of finite depth. *J. Geotech. Eng.*, ASCE, **111** (4), 479–499.

Savvidou, C. (1984) Heat Transfer and Consolidation in Saturated Soils. PhD. Thesis, Cambridge University.

Savvidou, C. (1988) Centrifuge modelling of heat transfer in soil. *Proc. Int. Conf. Geotechnical Centrifuge Modelling, Paris*, pp. 583–591. Balkema, Rotterdam.
Schowalter, W.R. (1965) Stability criteria for miscible displacement of fluids from a porous media. *A.I.Ch.E. J.*, **11** (1), 99–105.
Smedt, F. De and Wierenga, P.J. (1979) A generalised solution for solute flow in soils with mobile and immobile water. *Water Res. Res.*, **15** (5), 1137–1141.
Turner, J.S. (1973) *Buoyancy Effects in Fluids*. Cambridge University Press, Cambridge.
Wooding, R.A. (1959) The stability of a viscous liquid in a vertical tube containing porous material. *Proc. R. Soc. Lond.*, Series A, **252**, 120–134.

9 Cold regions' engineering
C.C. SMITH

9.1 Introduction

The cold regions of the world are areas where engineers must deal with climatic extremes and where human technology is often stretched to the limit. Huge forces can be involved (the iceberg scour of pipelines buried beneath the sea bed; the impact of wind driven ice sheets on drilling rigs; the jacking out of the ground of piles by frost heave), and yet the light touch of simply walking on the tundra can so radically change thermal boundary conditions that within years extensive thawing will have taken place, melting ground ice and resulting in major settlements.

Engineering in cold regions is a very expensive undertaking. Practical experience in many areas is still quite limited and while field tests have been and are being undertaken, the costs are often prohibitive, working conditions difficult or hazardous and the time scales very long if freezing or thawing behaviour is of interest.

Centrifuge modelling of such problems is an attractive proposition, with its ability to compress time scales, and to model large forces under controlled conditions. However, while it has already been used to study several cold regions' problems, this application is still very much in its infancy. Before it can achieve wide acceptance, there are fundamental questions regarding scaling to be answered, and technological developments to be achieved in order to permit controlled testing. It is the aim of this chapter to address both these issues and to point to some of the ways forward. Section 9.2 discusses the fundamental scaling laws pertinent to cold regions, and identifies those problems that should benefit from centrifuge modelling and those areas where fundamental knowledge is still lacking. The design of a model is discussed in section 9.3. Section 9.4 discusses the actual technology that is available, and section 9.5 considers package design. Section 9.6 gives some examples of specific tests, and concluding remarks are given in section 9.7.

9.2 Scaling laws and cold regions' processes

9.2.1 Introduction

This section briefly discusses those scaling laws relevant to cold regions already listed in the literature for general geotechnical and geo-environmental problems and then goes on to discuss in detail the modelling and scaling of specific cold regions processes. Further discussion can be found in Ketcham (1990) and Smith (1992).

9.2.2 Scaling laws

In discussing scaling laws it will be assumed that the reader is already familiar with the concepts of model and prototype. Cold regions' processes encompass both 'conventional' geotechnical considerations and processes similar to those encountered in the study of heat transfer and pollution migration.

Those scaling laws concerning the modelling of stresses, strains, seepage processes, consolidation, particle size effect, etc. have been discussed in connection with more conventional geotechnical problems by many authors (e.g. Schofield, 1980), while heat transfer by conduction and convection, density driven flow, solute migration due to concentration gradients and moisture transport in unsaturated soils have been discussed by Culligan-Hensley and Savvidou (chapter 8) and Smith (1992). These scaling laws are summarised in Table 9.1.

Since cold regions' problems are dominated by the properties of water and ice, and since properties such as viscosity, density, and creep rate vary significantly and non-linearly over small temperature ranges, it is generally necessary to impose the condition of similarity in temperature between model and prototype in order to achieve similarity of material properties. A similar requirement is required for the correct modelling of phase change.

From Table 9.1 it can be seen that consistent scaling of all listed parameters

Table 9.1 Centrifuge scaling laws

Physical quantity	Prototype	Model
Defined		
Macroscopic length	1	$1/N$
Microscopic length	1	1
Gravitational acceleration	1	N
Temperature	1	1
Derived		
Strain	1	1
Pore-water pressure	1	1
Stress	1	1
Time		
diffusion processes	1	$1/N^2$
inertial events	1	$1/N$
viscous processes	1	1
Total water potential	1	1
Interstitial water velocity	1	N
Moisture flux	1	N
Heat flux	1	N

is possible with the exception of the time scaling of inertial, diffusive and viscous processes, and the potential scale conflicts between microscopic and macroscopic lengths.

Diffusion processes encompass heat transfer by conduction and convection, density driven flow, seepage and consolidation. Inertial processes refer to dynamic events such as earthquakes, or explosions, while viscous processes relate to creep and to strain rate effects. The areas where the above mentioned scaling conflicts may affect specific cold regions processes are discussed in the following sub-sections.

9.2.3   Freezing processes

9.2.3.1   Sea ice.   The formation of sea ice is a complex event. It involves processes of freezing, heat conduction, and complex convection phenomena due to water having its maximum density at $\sim 4°C$. There is salt diffusion due to concentration gradients, salt expulsion from water upon freezing, and accompanying density driven flow. While little work has been done on the detailed dimensional analysis of the scaling of this process, there exist several reports in the literature (Langhorne and Robinson, 1983; Lovell and Schofield, 1986) where it has been experimentally demonstrated that the columnar structure of sea ice can be modelled successfully on the centrifuge with the ice grain size and structure reproduced in a scaled manner. Preliminary experimental evidence (indentation tests, Lovell and Schofield, 1986; crushing against a vertical pile, Clough *et al.*, 1986, Clough and Vinson, 1986) suggests that when this structure together with self-weight effects dominate the mechanical response of the ice, then ice–structure interaction models tests should yield correctly scaled results.

9.2.3.2   Frozen soil.   During the construction of centrifuge models it is normal practice to ensure that at a microscopic level the model soil is identical to the prototype soil in order that the stress–strain constitutive behaviour is the same in model and prototype. However, when a saturated soil freezes, the interstitial water undergoes a phase change to form polycrystalline ice, and as yet little is known about the effects of high acceleration on those changes. For sea ice, the ice grains so formed appear to be of a scaled down size (macroscopic length scaling). Should this be the case for frozen soil then the model frozen soil will not be identical to the equivalent prototype soil; and may thus respond differently (and perhaps benefically, see section 9.2.5.1) with regard to, for example, creep. If, however, the ice grain size is controlled exclusively by the pore sizes (microscopic length scaling), then it is reasonable to expect that the model frozen soil will behave similarly to the prototype soil. Resolution of this question is needed before tests can be correctly designed in which it is the behaviour of the frozen soil that is of interest. It is necessary to know whether a frozen soil model can be prepared in the laboratory and tested in flight or whether both the freezing and testing need to be done in flight in order to obtain the correct behaviour.

The formation of ice lenses during centrifuge model tests of frost heave in fine grained soils is another area which is poorly understood. It is generally accepted that ice lenses form within a volume of already frozen soil directly behind the freezing front and that at this position large suctions are generated which act to draw moisture out of the adjacent unfrozen soil through the already frozen material (the frozen fringe) to form an ice lens. This process can be said to be a function of stress level, temperature gradient, time and soil type, and will involve processes of heat conduction, phase change, migration of moisture through unfrozen and frozen soil, generation of suctions within the freezing front, salt migration in subsea soils, and creep in overlying soils. It would be expected that the migration of moisture through frozen and unsaturated soil should scale as that process for saturated soil as long as it is governed by a linear process akin to Darcy's law. In this case the 'permeability' will be a function of unfrozen moisture content and temperature (unless regelation plays a significant part in the transfer of moisture) which should be identical in model and prototype. If the generation of suctions is assumed to be purely a function of pressure, temperature and soil/water properties, which are identical in model and prototype then this should also scale 1:1.

The creep response of the frozen soil surrounding the area undergoing frost heave is important in that this frozen soil may restrict the growing ice lenses and thus provide a significant constraining pressure, the magnitude of which is dependent on that soil's ability to creep. Frost heave may thus be expected to scale consistently except where creep is a significant factor. An additional concern regarding the scaling concerns the width of the frozen fringe. It is unknown whether the width of this frozen fringe will scale as a macroscopic or a microscopic length in a centrifuge test where macroscopic thermal gradients will be N times larger (where N is the scale factor) but with the stress level the same as in prototype. If it scales as a microscopic length then this may lead to concerns similar to those made for particle size effects in conventional tests. To date, very little work on the centrifuge scaling of frost heave has been published in the literature. A preliminary modelling of models study by Chen et al. (1993) on one-dimensional closed-system frost heave would appear to provide evidence for the correct scaling of frost heave in sand and clay although more data are required to confirm this with confidence.

9.2.4 Thawing processes

A thaw settlement problem will be dominated by processes subject to diffusion time scaling such as heat transfer due to conduction and convection, thawing, generation of excess pore pressures at the thaw front, pore water flow and consolidation effects (Smith, 1992). Due to the volume reduction of ice on thawing to water, some zones of an originally ice-saturated model may also become partially saturated. The problem may thus also involve processes of heat and moisture transport in unsaturated soil. Where thaw bowl growth takes place in sub-sea permafrost, salt migration will also be a factor and Palmer *et al.* (1985) indicate that this should also be subject to diffusion time scaling. In general all processes occurring during thaw settlement should scale consistently in a centrifuge model, with the exception of situations where the deformation of frozen soil surrounding or enclosed by a thaw zone is significant; creep time scaling could give rise to time scale discrepancies.

9.2.5   Mechanical response of frozen soil and ice

The mechanical response of ice and frozen soil, encompasses such processes as creep, crushing, buckling, and crack propagation, as well as elasto-plastic deformation and shear failure.

9.2.5.1   Creep.   Creep of frozen soil can involve several temperature and stress-dependent mechanisms such as pressure melting, regelation, crystal reorientation, grain growth (diffusion creep) and dislocation (power law) creep. Each will operate at different rates and these rates may be subject to different scalings. The time scaling of creep is therefore difficult to quantify in the general case, though in specific cases one mechanism may dominate the others.

While it is possible that the time scale for the creep of frozen soils may vary between 1 and N^2, it is generally assumed that since stresses and temperatures are the same in model and prototype, the time scaling of creep events in model and prototype will also be identical, but this has not been experimentally demonstrated. Certain considerations may lead to different expectations. If the ice grain size is reduced in a centrifuge model then this will tend to accelerate the process of diffusion (low stress, linear viscous) creep. A grain size reduction of N in a pure material may result in a creep acceleration of N^2 (Ashby and Jones, 1980). In addition, creep effects resulting from the redistribution of unfrozen water in frozen soils due to thermal and stress gradients may in fact scale consistently with other diffusion processes such as heat conduction.

In circumstances where creep time scaling is identical, time scale conflicts will arise where inertial or diffusive processes are also significant, preventing the achievement of complete similarity, and model tests may be restricted to those where this is the only process, e.g. load response of frozen ground. In addition, where a 1:1 scaling occurs, a major time compression benefit of the centrifuge is lost. Model tests become attractive only when they are of a short-term nature. Possible solutions are to use different soils, or soil or ice at a warmer temperature to model the field situation such that the model creeps faster. Basic research is needed to show that the creep response can be correctly represented this way. Tests of such nature will, of course, still represent a geotechnical event useful for the calibration of numerical and analytical models.

One of the major benefits of centrifuge modelling of unfrozen soil is the correct simulation of *in situ* stresses and thus the stress–strain response of the soil. However, in contrast to unfrozen soil, the strength of frozen ground is often dominated by the cohesive strength of the interparticle ice which often greatly exceeds the magnitude of the overburden stresses. The self-weight benefit of the centrifuge is thus often regarded as negligible, though self-weight may be significant for problems involving for example tunnels, cavities, or subsurface thawing (Smith, 1992) in frozen soil. Where creep enters the tertiary failure stage, the benefit of centrifuge modelling becomes more significant. If frozen soil is regarded as a frictional material held together by a creeping matrix of ice crystals, ultimate strength will be as much stress dependent as for unfrozen soil, and model tests where frozen soil is loaded to failure may work well in a centrifuge.

9.2.5.2 Crushing, buckling and crack propagation. The strength of frozen soils and ice, and their mode of failure (ductile or brittle) in response to loading is strongly affected by strain rate. Similarly, crushing, buckling and crack propagation, normally associated with the deformation of ice sheets, are also strain rate determined. It is thus important that where these processes are being modelled, strain rates are similar to the prototype event, conflicting with inertial and diffusion time scales. However, it is likely that many experiments involving ice sheets may progress in two stages: the growth of the ice sheet, followed by its deformation, with each stage being subject to a different scaling.

Palmer (1991) in a theoretical study of the centrifuge scaling of crack propagation in brittle materials indicates that in circumstances where cracks scale as the system as a whole, crack stress intensities will not be correctly modelled and will scale as indicated in Table 9.2. This led Palmer to the conclusion that, for correct modelling of crack propagation, the acceleration scale factor ought to be the 3/2 power of the length scale factor, i.e. a 1/100 scale model should be loaded by a 1000 g acceleration. This may mean that

Table 9.2 Centrifuge scaling of crack propagation

Physical quantity	Prototype	Model
Assumed		
Crack length	1	$1/N$
Derived		
Crack stress intensity	1	$1/N^{1/2}$

frozen models need to be formed at one acceleration level, but tested at another.

9.2.5.3 Dynamic loading. Dynamic events in cold regions include earthquakes and icequakes. The latter events may involve the pressing of ice sheets on one another or upon an artificial island or caisson and if inertial time scaling is operative, this will be in conflict with the scaling of diffusion, creep and strain rate processes. The dynamic testing of models consisting of both frozen and unfrozen soil or, for example, of ice sheets impacting a soil filled caisson will thus require careful design and interpretation.

9.2.6 Summary

A survey of scaling laws and experimental evidence indicates generally good prospects for modelling cold regions events, especially for problems involving thawing, sea ice growth, or those in which restraint of, or loading from, surrounding unfrozen soil is significant. Those processes for which there are doubts or inconsistencies are those in which simultaneous creep, inertial effects, diffusion processes and strain rate-determined processes are occurring, or in which brittle fracture plays a significant part. Many tests will thus be restricted to a regime where one process dominates, or to stages where one process dominates in the first stage and another in the second.

   More basic research on scaling needs to be done before the benefit of centrifuge modelling can be brought to many frozen soil problems. If these uncertainties can be resolved, there is much to be gained from the centrifuge modelling of frozen soil. The fundamental studies which are required must elucidate the scaling of creep response, frozen fringe magnitude, ice crystal size in frozen soil, crack propagation in frozen material, and the effect of freezing in laboratory models or in centrifuge flight on the structure and constitutive response of a frozen soil.

9.3 Modelling design considerations

9.3.1 Introduction

This section will consider a typical cold regions' test, the required boundary conditions, the degree of heat transfer needed, and the modelling methodology required to implement a test. The simplest design methodology is to idealise the field situation to such an extent that it can be modelled at normal temperatures, while still retaining the aspect of interest in the original problem. Much of the initial cold regions' work carried out on the centrifuge was performed this way: pipeline thaw settlement (Vinson and Palmer, 1988; Phillips, 1985), caisson subjected to ice-quake (Jeyatharan, 1992), iceberg scour (Lach *et al.*, 1993). As long as fundamental processes occurring in frozen material are not neglected then this can prove a useful and straightforward technique for modelling cold regions' events. However the purpose of this section is to discuss models in which frozen material is being tested and in which heat transfer is required during centrifuge flight.

9.3.2   Classification of cold regions' tests by heat transfer

Consideration of the class of cold regions' centrifuge tests that might be modelled leads to the conclusion that there are two distinct types, each requiring a correspondingly different magnitude of heat extraction or input.

1. Tests involving small heat input/output requirements in flight; these include tests that will typically involve pre-frozen models undergoing thawing or creep deformation. Requirements may thus include:
   (a) maintenance of a constant temperature in the model (overcoming insulation losses),
   (b) maintenance of a small geo-thermal gradient, and
   (c) extraction of small quantities of heat generated by a structure within the model.
   These processes might typically require approximately 100 W of cooling/ heating for a typical medium sized centrifuge package.
2. Tests involving large heat input/output requirements; such tests may involve:
   (a) Full freeze–thaw modelling of permafrost occurring over a simulated 'centrifuge year'. This involves extraction and input of large amounts of cold/heat which is absorbed by the large latent heat capacity of the soil water.
   (b) Freezing of sea or fresh water to form ice, for subsequent testing in, for example, indentation or impact tests.
   The magnitude of the rates of heat transfer required in such centrifuge models may be readily appreciated through the consideration that during a moderate annual freeze–thaw cycle of soil, where, for example, the surface temperature varies from $-10°C$ in winter to $+10°C$ in summer, the soil receives or emits a peak heat flux of around $30 \, W/m^2$. When it is further considered that heat flux in a centrifuge model should be N times larger than in the prototype case, where N is the scale factor (see Table 9.1), it is seen that extremely high flux rates are required for even relatively low g tests, for example $1.5 \, kW/m^2$ in a 1:50th scale test conducted at $50 \, g$. Coupled with the fact that this heat must be transfered through the air to or from a free soil surface, it is seen that this raises one of the biggest

challenges in the design of centrifuge equipment to achieve truly scaled model freeze–thaw cycles.

It may of course be unnecessary to achieve correctly scaled freezing rates. A frozen soil or ice sheet will still result regardless of the rate of freezing, and an imperfect model may, in some cases, be better than nothing.

9.3.3 Temperature ranges

Consideration of the range of temperatures experienced in the cold regions of the Earth gives an idea of the range of temperatures that might be required of centrifuge models. The record low air temperature recorded in Canada has been reported as $-63°C$. However it is the temperature of the soil or ice under study that is of interest and fractionally below the surface this may be significantly warmer than the air. A more reasonable minimum temperature of interest would be about $-30°C$. If this is the lowest temperature of the soil or ice from which heat must be transfered to cause freezing, then the source of this cold must be at a significantly lower temperature if the rates of heat transfer are to be of the order discussed above. It may therefore be necessary to use air/gas temperatures as low as perhaps $-100°C$. This is feasible, but with such low temperatures, problems may begin to arise with the temperature effects on containment systems and ability to satisfactorily control heat-transfer rates.

Artificial ground freezing, which may take place at temperatures down to around $-40°C$, is less of a concern since this normally takes place by direct conduction to the soil via refrigerating pipes, and so the model pipe temperatures need only match the prototype temperatures. Where prototype ground freezing is performed using liquid nitrogen, then clearly similar temperatures will need to be used within a model, requiring the use of liquid nitrogen or a very cold gas.

9.3.4 Boundary conditions

The requirements of a large proportion of applications will be satisfied by the temperature control of the model base and of the air above the model, with the model sides maintained as adiabatic. Surface temperature control is almost essential, as this is where the dominant heat transfer occurs in the field. Control of the base temperature is desirable for two main reasons. It permits the modelling of the natural geothermal gradient found in the earth, and it can also assist with the thermal pre-conditioning of a model prior to testing, in that it is much easier to extract heat by direct conduction through the base than indirectly through the surface. Thus, for example, where a model is to be pre-cooled in flight to just above 0°C, prior to freezing from the surface, this can be much more rapidly and efficiently achieved by cooling from below (though in clays, this might result in undesirable thermally induced pore pressure effects).

If uniformity of heat transfer can be achieved, uniform conditions or depth of freezing or thawing can be maintained and relied upon. Accurate control of these boundary conditions is desirable and significantly influences the choice of freezing technology.

9.3.5 Preparation of frozen soil

In modelling frozen ground, it is necessary to determine whether this can be frozen during model preparation at $1g$ on the lab floor and then tested in flight or whether it is necessary to freeze the soil in flight. The former minimises flight time; however as discussed in section 9.2.3.2, freezing in flight may have a significant effect on the resulting frozen soil structure. The choice depends on which properties are of greatest interest during the test. For example, where a test is concerned with thaw settlement in a predominately granular soil, and in particular with the stress–strain response of the thawing soil rather than the frozen soil, then it may be reasonable to prepare the model beforehand and thaw in flight (e.g. Smith, 1992).

9.3.6 Modelling of uncoupled processes

Where several processes occur in the problem, but are uncoupled, it may be desirable not to model accurately one process in order to maintain model simplicity. This will be true for the case of, for example, iceberg scour over a fine-grained soil where all soil behaviour will be essentially undrained for velocities above a certain critical value and in this case the velocity is essentially uncoupled from the mechanical processes in the soil (up to the point at which inertial or strain rate effects are introduced; Lach et al., 1993). Similarly where thawing takes place in a coarse grained soil, the stress–strain response of the soil will be essentially uncoupled from the thermal processes for thaw rates below a certain (high) value.

9.3.7 Modelling of models

If models of different scales give test results which, when scaled up, predict the same behaviour of a full-scale prototype, then this is convincing proof of the correctness of the scaling principles and of the model test procedures. A set of models (Malushitsky, 1981) of a mine waste heap performed at scales of $N = 186.9$, $N = 148.9$, $N = 124.3$ appears typical of what has been called 'the model testing of the models' in Russian model test practice. Few such tests of cold regions problems have been reported, but it would, in principle, be a way to check the scaling principles of cold regions' tests.

9.4 Equipment

9.4.1 Introduction

Having determined the nature of an experiment, the interpretation of the scaling and the nature of the boundary conditions, it is necessary to consider how this experiment may be implemented as a centrifuge test and the range of equipment and techniques that are available to achieve this. There is clearly a need to provide a cold source in order to extract heat at low temperatures from a package on the arm. This may be done by passing a cold fluid through slip rings or by securing a cold source on the arm or within the package itself. It is important to discuss several different techniques for achieving this since any one application will have special requirements best provided by one particular cold source.

The performance of any cooling system will be a function of two separate aspects:

1. Its ability to extract heat from the model at the desired rate and at the desired temperature.
2. Its ability to absorb this heat for the duration of the test.

The former is a function of the heat-transfer method as described in section 9.4.4 and is strongly dependent on the temperature of the cold source. The latter is a function of the chosen cooling system (section 9.4.2). In general, the enhancement of one will tend to diminish the other and there will be an optimum balance between the two.

Consideration also needs to be made of the specification of new slip rings and technologies that can be used with older existing slip rings of perhaps a lower specification. The cooling of a free soil or water surface will occur predominantly by convection (whether forced or natural) and a very much lower air temperature is required than the eventual desired soil/ice temperature in order to attain useful heat transfer rates. Additionally, where freeze–thaw processes are of interest, it will also be necessary to include a source of heating within the package; this may be an independent system or in certain circumstances a reconfiguration of the cooling system.

9.4.2 Cold sources

This section considers a range of possible cooling systems that can be used with a centrifuge. The emphasis is on reliable systems with a minimum of moving parts. Each will be considered in turn as follows together with its advantages and disadvantages. References are given where this type of system is in existence or has been discussed previously in the literature.

1. Circulation of an externally cooled liquid refrigerant through slip rings, along the centrifuge arm, and out to the package. Possible secondary

refrigerants include alcohol, glycol, silicone oil or brine (Jessberger and Güttler, 1988).

Advantages: All cooling done external to arm (requires refrigeration plant). Potentially large continuous power capacity.

Disadvantages: Inevitable heat loss on passing the slip ring and the arm necessitating insulation. Fluid return requirement results in a high pressure at the package giving rise to potential safety hazards if leaks develop. High-pressure, low-temperature fittings required. May require fitting of new slip rings. Safety problem with very cold refrigerant leaks which could cause damage to the centrifuge itself.

2. On arm refrigeration unit, circulating liquid secondary refrigerant (as above) to package.

   Advantages: Potentially large continuous power delivery. No slip-ring changes.

   Disadvantages: High head of fluid at package. Safety hazard if leaks develop. High-pressure fittings required. Safety hazard if refrigeration plant fails. Specialised refrigeration plant design required for high g-level operation. Reliability may be a problem. Large unit required. Balancing necessary.

3. Expansion of high pressure gas through a vortex tube (Otten, 1958) to achieve cooling.

   Advantages: Uses existing slip rings and used air is vented into the centrifuge pit. Small unit. One-way circulation of air does not give rise to high pressures. Moderate continuous power delivery. Useful for direct air-cooling applications. (Also generates warm air stream, useful for heating applications.) No moving parts. Safe. Cold is generated in the package— no loss of heat in slip rings or line insulation required. For enhanced cooling it can be supplied with pre-chilled gas.

   Disadvantages: High flow rates of air required to achieve necessary cooling (air may require lubrication to pass through slip ring and then be filtered for the vortex tube). Gas used needs to be dry to prevent clogging of vortex tube outlet with ice particles; either use a dehumidifier in conjunction with the compressor, use bottled gas, or inject an antifreeze mist into the air stream.

4. Cooling using solid-state Peltier effect heat pumps.

   Advantages: Precise electronic temperature control of localised areas. Can typically achieve up to a 60°C temperature differential. No moving parts. Small device. Can be used as a heater by reversing direction of current flow. Continuous operation possible.

   Disadvantages: Generation of a high temperature differential results in a lower power transmission. Operation of device generates additional Joule heating which needs to be dissipated (typically by means of a liquid coolant); this becomes very large for a high heat pumping rate, and requires high additional cooling. Careful mounting in the centrifuge acceleration field required to eliminate large stresses on this relatively fragile device. Specialised d.c. power supply requirements—low voltage, high current—needs either on-board power supply or slip rings capable of transmitting high currents.

5. Transmission of a liquefied or very cold gas through the slip rings to be vented at the package (no recirculation). Possible coolants are liquid CO_2, liquid N_2 or cold N_2 gas (Jessberger and Güttler, 1988; Schofield and Taylor, 1988).

   Advantages: High continuous power transmission as a liquid. Can either be circulated as a liquid through the model to achieve freezing or can be expanded to provide a cold gas, or sprayed as droplets giving evaporative cooling. Very low temperatures achievable. One-way circulation.

   Disadvantages: Liquid CO_2 or N_2 source required. Liquid N_2 is very cold requiring great care if contact with temperature sensitive load-bearing components is possible. Danger of oxygen condensation from the atmosphere (Lovell and Schofield, 1986). Liquid CO_2 must be circulated at a high pressure but has a less hazardous temperature. If taken to the package as a liquid then high-pressure fittings required. Specialist slip ring system required.

6. Cooling using liquid N_2, liquid or solid CO_2 stored on the package or arm (Lovell and Schofield, 1986).

   Advantages: Very low temperatures attainable, high heat-transfer rates. No slip-ring concerns. One-way circulation with used cold gas vented into the centrifuge pit.

   Disadvantages: Liquid N_2 can be difficult to handle, store and control, and spillage onto structural steel could lead to brittle fracture. Limited source of cold. (Other disadvantages as above.)

It is not possible to recommend any particular technique or combination of techniques, as this will be determined by the nature of the application.

9.4.3   Heat sources

As with cooling systems there exist several choices that can be made for the heating system. These can be listed as follows.

1. Expansion of high-pressure air/nitrogen through a vortex tube to produce a hot gas stream.

   Advantages: As for previous consideration (section 9.4.2). Also generates cold air stream—useful for cooling applications.

   Disadvantages: As for previous consideration.

2. Electrical heating of a pumped air stream.

   Advantages: With feedback, can give good control of temperature. Can be used in conjunction with a cold air stream providing a wide range of temperatures.

Disadvantages: Requires air supply through slip ring, or fan arrangement on arm or package.

3. Direct electrical resistance heating.

   Advantages: With feedback, can give good localised control of temperature. Robust, flexible.

   Disadvantages: Cannot be used to cool, therefore must be used in addition to a cooling system.

4. Heating using solid-state Peltier effect heat pumps.

   Advantages: Precise electronic temperature control of localised areas. Heating comes not only from Joule heating but also by being pumped from the reverse face. No moving parts. Small device. Can be used for cooling in reverse mode.

   Disadvantages: Careful mounting required to eliminate large stresses on this relatively fragile device.

9.4.4 Heat and cold transfer to the model

9.4.4.1 Introduction. This section will consider the methods for transferring heat energy from a cold/heat source to or from the model itself. This is largely independent of the actual source of cooling or heating chosen and will involve heat transfer to the model surface and the model base. This section is most relevant to a system requiring large heat input/extraction. A system requiring the maintenance of a constant temperature is more straightforward and requires less detailed attention.

9.4.4.2 Transfer of energy to the free surface of a model. The transfer of heat to and from a model free surface is complicated by the necessity of having an air gap above the model surface so as to maintain a zero stress condition at zero depth and to permit unrestrained surface movement. Heat may be transferred either directly or indirectly as follows.

Direct heat transfer. This could involve the use of a cold gas stream or 'wind' blowing across the model to provide a rapid method of cooling using forced convection. There are several associated problems as follows:

1. As the cold gas passes over the surface it warms up, causing uneven cooling of the model surface. This will occur especially in circumstances where a high heat-transfer rate is needed.

2. Regardless of whether the gas circulated is in an open or closed loop, moisture may be transferred from the model surface and deposited elsewhere in the cooling system.

3. A cold gas generator is required.

4. The velocity of the required gas stream may be such as to damage the model.

For a correct arrangement of gas flow pattern over the model surface to give evenness of the heat transfer, there are several options:

1. Cold gas can be blown laterally across the model, perhaps using a graded system of baffles to create turbulence and to spread the gas flow (and so the heat transfer) evenly across the model.
2. Cold gas may be injected downwards through a central hole above an axi-symmetric model and vented through holes at the sides or vice versa (for example, Lovell and Schofield, 1986).
3. Gas can be injected at several locations and then vented at intermediate positions.

An alternative system for use in situations where extremely high cooling rates are required could involve the use of evaporative cooling. This may involve, for example, the spraying of liquid nitrogen droplets or dry ice particles onto the model surface which would then evaporate.

Indirect heat transfer. This involves the mounting of a heat-exchanger plate a small distance above the model surface and relies on conduction, convection and radiation heat transfer through the air to transfer heat between it and the model. If a uniform plate temperature is maintained, this gives uniform heat transfer across the entire model surface and maintains an enclosed air volume above the model surface. The system will be independent of the cooling source chosen, but the presence of the required plate may limit the addition of extra instrumentation, and the air volume between model and plate must be well sealed from external influences.

Appendix A details the computation of heat transfer between plate and model surface. For example, a cooling plate maintained at $-50°C$ will transfer $\sim 750\,W/m^2$ to a model surface at $0°C$ at a centrifuge acceleration of $50\,g$, and the heat transfer will be virtually independent of the plate/model surface separation. The system has low heat-transfer capacity at low model accelerations, due to reduced convection. However, in simulating any given prototype situation, the heat flux to be modelled will scale directly as the model scale factor and at lower accelerations, a lower heat flux will be required which will tend to counteract the reduction in heat-transfer capability.

In circumstances where the model needs to be warmed up, the transfer of heat from a warm plate to a cooler model surface will involve minimal convection and so the same rates must be achieved with a higher plate temperature to maximise radiative transfer. The generation of heat is a more straightforward process than the generation of cold; however, high temperatures could cause problems with surrounding instrumentation or package walls. Figure 9.1 shows the predicted equivalent prototype heat-transfer rates achievable using a natural convection system, as a function of temperature difference and centrifuge acceleration.

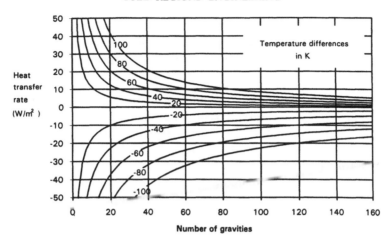

Figure 9.1 Plot of prototype heat transfer rates achievable as a function of temperature difference and centrifuge acceleration.

If an air gap above the model surface is not essential, then filling it with lightweight shredded aluminium foil would markedly improve the heat-transfer capabilities. One other alternative is the use of thin lightweight heating mats placed on the surface of the model. These could be used to achieve fine thermal control of the surface temperature, but would still require a cold atmosphere above for freezing and could affect the zero stress surface condition.

9.4.4.3 Transfer of energy to the model base. The absence of an air gap makes transfer of heat to the model base much less of a problem than to the surface. Direct conduction is possible and while very low temperatures are not required, the supply of water to the model base via a drain must be considered.

9.4.4.4 Heat exchange within the model. It may be desirable to model such structures as frost heaving pipelines, liquefied gas tanks, or processes such as shaft freezing within a soil model for which a separate feed of coolant would be required. The connection of a gas coolant to the structure is most straightforward mechanically, but consideration needs to be made of the possibly large loss of temperature occurring as the gas flows through the structure. In contrast, a liquid refrigerant will lose much less temperature for the same heat extraction rate but is more difficult to connect as it is likely to be under a very high pressure in centrifuge flight. Fittings will be much less flexible, which may affect the structure if it is to move freely. A liquid is also more likely to affect instrumentation internal to the structure such as strain gauges.

9.4.5 Slip-ring specifications

The discussion of possible cold sources above included a number of possibilities that may necessitate specialist slip-ring requirements. These are summarised below.

1. A pair of slip rings capable of circulating a low-temperature refrigerant. The pipes down the arm should be insulated and fittings at the end capable of sustaining the head of fluid generated down the arm at high g levels. The slip rings should also be insulated or separated from the other rings.
2. One or two slip rings and associated insulated lines able to take liquid carbon dioxide. This can be circulated at a high pressure and room temperature or at a low pressure and low temperature, e.g. 6.6 MPa (460 psi) at 27°C or 1.2 MPa (90 psi) at −33°C.
3. One slip ring able to take high flow rates of compressed gas preferably to at least 3.5 MPa (250 psi) for use with vortex tubes and other compressed gas equipment. The larger the slip ring, the less friction heating will occur. Cooling of this ring or separation from other rings may need to be considered. The line down the arm may need insulating and be provided with heat exchanging facilities. It may be beneficial to insulate the slip rings if it is desired to circulate already cold gas through them; unless, however, the rings are designed for negligible friction heating, this would not be worthwhile.
4. A pair of electrical slip rings able to transmit high d.c. currents for use with Peltier devices. An on arm transformer and control package could be used instead, where there is a large area of free space on the arm itself.
5. A vacuum insulated line capable of carrying liquid nitrogen (Jessberger and Güttler, 1988).

9.4.6 Safety

Safety is paramount in a centrifuge model test and consideration must be made of the consequences of a failure in control of an electrical system, or of a mechanical component. Several different aspects can be considered as follows.

1. *Failure of temperature sensors.* This could lead to overchilling or overheating of the package. Overchilling is less likely to be a problem as the minimum temperature achievable within the system should be known. However heat gain could continue to very high temperatures in an insulated package. Overheating could lead to the risk of gross creep deformation of any plastic component such as a Perspex window. The melting of insulation and subsequent short circuiting could lead to a fire risk and damage to other wiring in the centrifuge system as a whole and to the slip-ring stack. Heating a secondary refrigerant that is contained in a closed circuit could lead to overpressurisation, and overheating a sealed

model could lead to generation of pressurised steam. Most of these effects can be mitigated through protection of electrical circuits with fuses, installation of an independent temperature sensor anywhere where heating is occurring and monitoring of the package load.

2. *Leakage of refrigerant.* A liquid refrigerant will generally be under high pressure during a centrifuge test. Leakage will result in spraying of the fluid over the package or the centrifuge. This may be problematic if the refrigerant chosen is corrosive or toxic. For this reason, a relatively benign refrigerant such as brine or silicone oil is preferable. If the refrigerant is being used to cool devices that generate heat, then possible overheating as discussed above may occur. Leakage of liquid nitrogen on to a load-bearing component may result in failure due to brittle fracture. Passage of liquid nitrogen through pipes in the atmosphere may also result in condensation of oxygen outside the pipes. If this comes into contact with hydrocarbon material, there is a possible risk of explosion.

3. *Failure of slip ring lubrication.* If large volumes of compressed air are being supplied to the package through a slip ring that needs lubrication, then failure of the lubrication system may result in the drying out and subsequent damage to the slip-ring seals. The lubrication system therefore needs to be regularly checked.

4. *Temperature effects on mechanical components.* Heating will accelerate creep processes especially in plastics, and thermal cycling may result in fatigue effects on highly stressed components, which should be periodically checked. Chilling will tend to increase the brittleness especially of plastics and again this needs to be considered. Metals such as copper, stainless steel and aluminium alloy generally suffer no weakening effects down to very low temperatures.

5. *Freezing of drains.* Freezing of the base drain is only likely to affect the actual model test in progress. However, it is important that no leaks occur so that no enclosed volumes of water freeze since due to the 9% expansion of water in its transition to ice, the large pressures generated could be sufficient to split certain fittings.

9.5 Package design and ancillary equipment

9.5.1 Package design

The complexity of cold regions' centrifuge tests is mainly concerned with the need to control thermal conditions within the model. Experience has shown that four main phases may need to be considered which will affect package design.

1. Model preparation.

2. Pre-test thermal conditioning.
3. Test activity.
4. Post-test investigation.

Test models to be frozen in flight may be prepared at temperatures just above freezing to minimise the freezing time in flight. Similarly, thawing tests may require the model to be pre-frozen before testing. In flight, it is necessary to maintain appropriate boundary conditions (both thermal and displacement) and to insulate the package to minimise the gain or loss of heat. During the test a non-equilibrium thermal regime may be set up within the model, e.g. a collar of ice around a frost heaving pipe, which it is desirable to examine as rapidly as possible after the test. It is necessary to be able to dismantle the rig quickly and perhaps to be able to take radiographs of the model, requiring careful choice of container size and materials.

The basic concepts behind all cold regions' package designs to date have followed a broadly similar pattern. For safety a relatively large aluminium alloy/steel strongbox is used to contain the entire experiment. Within this box is placed a smaller inner container whose sides are constructed from an insulating material. This must be either thick enough to take all stresses envisaged in centrifuge flight, or be rigidly supported by packing the space between it and the strong box with stiff wood or plastic blocks that will also provide insulation. It is useful to construct one of the package sides from a transparent insulating material such as acrylic in order to be able to observe the model in flight. The thickness of the insulation will depend on the ratio of model thermal conductivity to insulation conductivity. During freezing processes, it may be necessary to account for large lateral stresses induced by the expansion of water during its transition to ice.

In contrast to the sides the base must be strong enough to take the weight of the model or package, be a conductor of heat where thermal control of the model base is necessary, and must provide drainage holes for water at the base. In order to provide easy access to the model after a test, it is useful to be able to detach the base from the model box sides so that the model can be easily extruded and sectioned.

The lid to the model will usually contain the surface freezing equipment and may also be used to mount instrumentation. Insulation may be provided by lightweight foam sheets. It is often useful to mount this equipment upon the inner box rather than the outer to simplify model assembly.

Alternative set-ups that have been used are a strongbox with external insulation attached (Clough et al., 1980), or a copper tank with insulating foam cast between it and a strong box (Lovell and Schofield, 1986). These experiments were used to investigate sea ice formation and were less concerned with lateral thermal boundary conditions. There was little necessity for post-test access to the model, or for detailed pre-test construction.

If model packages are designed with no removable inner containers and if pre-freezing or pre-cooling is necessary, then the entire package will need to be placed within a suitable cold room or cold cabinet. If pre-freezing involves very low temperatures, then there may be concern over thermal strains in the outer strong box, though it may be that the model can be pre-cooled by running any built in refrigeration systems in the laboratory.

Any package used to construct models from consolidated clay should be able to be used in conjunction with a consolidometer.

9.5.2 Ancillary equipment

Cold regions centrifuge experiments may require ancillary items such as:

1. Cold rooms or cold cabinets for model preparation, thermal conditioning and post-test model investigation. These may be run well below zero for the pre-freezing of models and the examination of frozen material after the tests, and also at just above zero for pre-conditioning a model to be frozen in flight.
2. Refrigeration system and pump for circulating refrigerant through the slip rings to the package.
3. A compressor and air dryer for supplying compressed air to a vortex tube.
4. Supply of liquid carbon dioxide or liquid nitrogen for circulating to the package, either through a slip ring or from an on-arm cylinder.
5. Source of solid carbon dioxide for use in pre-chilling gas on the arm.
6. Facilities for hooking up the package to refrigerant/gas supplies in the laboratory for 1 g testing or pre/post-conditioning of models.

9.6   Examples of tests and results

9.6.1   Introduction

Only a few actual cold regions' tests involving in-flight freezing, thawing or deformation of frozen material have been reported. These include sea ice (Langhorne and Robinson, 1983; Vinson and Wurst, 1985; Clough et al., 1986; Clough and Vinson, 1986; Lovell and Schofield, 1986; Jessberger, personal communication, 1993), creep of frozen ground (Jessberger, 1989), thaw settlement (Smith, 1992) and frost heave (Chen and Smith, 1993; Chen et al., 1993; Jessberger, personal communication, 1993).

It is the purpose of this chapter to give a flavour of the techniques used by some of these researchers to implement their tests with emphasis on the equipment used rather than the results. Where relevant the results have been discussed in previous sections.

9.6.2   Sea ice growth and indentation

Lovell and Schofield (1986) describe a series of model tests in which a sheet of

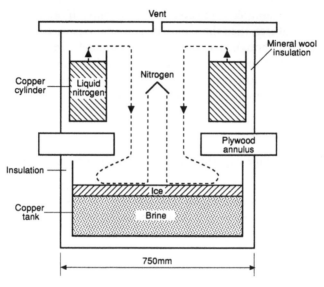

Figure 9.2 Sea ice package.

sea ice was grown in centrifuge flight and then subjected to a vertical indentation. The experimental rig consisted of a 850 mm diameter tub internally lined with foam insulation and containing a cylindrical copper tank. On top of the tub was an annular plywood plate supporting what was called a 'cold gas generator', which was a set of eight copper cylinders insulated with mineral wool. A second plywood plate was then placed above these to form a lid (Figure 9.2). Prior to the test, liquid nitrogen was placed within the copper cylinders and chilled brine was loaded into the copper tank. In flight, as the liquid nitrogen boiled, cold gas overflowed through the hole in the annular plate, fell onto the water surface and then rose back up the centre of the hole as indicated in the diagram. The level of the liquid nitrogen was indicated by a balsa wood float in each cylinder with a level stick protruding through the lid. As the liquid nitrogen evaporated and cold gas was generated, the levels were visible by CCTV. The ice formed as the cold gas flowed across the sea water was reported to have scaled-down grain-size structure similar to full-scale sea ice and the indentation tests indicated similar scaled strength of the ice sheet as reported for the prototype material. It was found that 50 litres of liquid nitrogen was sufficient to form an ice sheet 30 mm thick by 750 mm diameter in 4.5 hours.

9.6.3 Thaw settlement of pipelines

Smith (1992) reports a series of both plane strain and three-dimensional tests concerned with the modelling of a warm Arctic pipeline passing across a

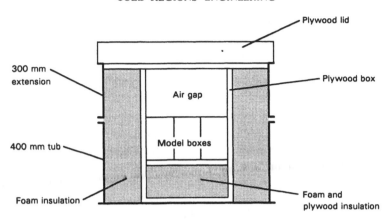

Figure 9.3 Insulation and containment for thaw settlement models.

transition from thaw stable to thaw unstable soil. Thaw-stable soil was modelled using saturated frozen fine sand, thaw-unstable soil using ice rich sand, and the active layer (where present) was modelled using dry sand. The models were constructed and pre-frozen in plywood containers in a cold cabinet and all instrumentation was attached at this stage. Five of these pre-frozen models were then loaded into an internally foam and wood insulated steel strong box and packed in tightly using wooden spacers. A wooden lid was then attached to the box (see Figure 9.3). In-flight the model pipelines, which contained heating elements, were energised and the response of the pipelines monitored. Emphasis was made on monitoring the down drag forces upon a rigid plane strain pipeline and investigating the deformation of a flexible aluminium alloy pipeline in the three-dimensional test (Figure 9.4). After the test individual models could be rapidly removed from the package in

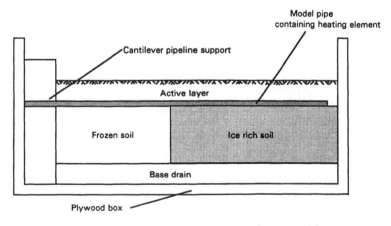

Figure 9.4 Three-dimensional thaw settlement model.

their containers, radiographs taken and then the models refrozen prior to sectioning. Results indicated that the pattern of thaw bowl growth had a significant influence on the forces acting upon the pipeline.

9.6.4   Pipeline frost heave

Chen and Smith (1993) describe a test designed to demonstrate the modelling of pipeline frost heave. An aluminium model pipeline of correctly scaled stiffness was embedded in a soil model such that it lay half in a body of sand and half in a body of silt, the latter being more frost susceptible than the former. The model was constructed in a container with an aluminium alloy base plate and transparent Perspex sides. This container was then placed in the centre of an aluminium alloy strongbox and insulated from the box using either plywood blocks or a Perspex block where a window was required (see Figure 9.5). Control of the water level in the model was via a standpipe system connected through to the aluminium alloy base plate of the inner box. Suspended above the soil model was a black plated aluminium alloy plate which was cooled by Peltier devices in turn cooled using a water system. This was to assist in maintaining a temperature of 0°C in the air above the model. On this plate were also mounted LVDTs whose spindles connected with rods sticking up from the pipeline so that heave could be monitored in flight. An insulating foam sheet and steel lid were then placed above the plate. On top of the lid was placed a vortex tube. This was fed compressed air through the slip rings and the output cold air stream was blown through the model pipe, resulting in freezing around the pipe and heave. Cold gas was the preferable coolant for this experiment as the pressure of a liquid refrigerant in the model pipe would make construction of the model to allow unrestrained movement of the pipe very difficult. (Successful pipeline frost heave tests have, however,

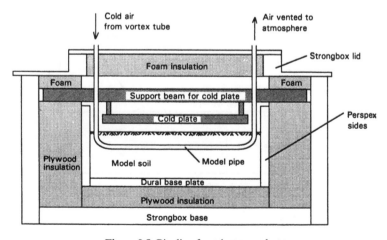

Figure 9.5 Pipeline frost heave package.

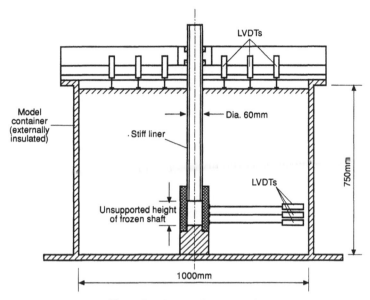

Figure 9.6 Ground freezing package.

been performed at the Bochum centrifuge using a circulated brine refrigerant; Jessberger, personal communication, 1993.)

The cooling water used for the Peltier devices was not recirculated but was sprayed out through a nozzle on the base of the package against the centrifuge pit wall. Here it evaporated and was carried away in the air flow. This had the added advantage of helping to keep the centrifuge pit cool. A nozzle was used in order to maintain some back pressure within the water feed line so as to minimise the effects of gas bubble formation due to cavitation.

9.6.5   Creep of artificially frozen ground

Jessberger (1989) reported experiments performed to study design problems associated with vertical shafts produced by artificial ground freezing. The study involved freezing of clay using a single vertical freeze pipe in the laboratory and an investigation of the subsequent creep behaviour of the frozen annulus of soil under centrifuge conditions using LVDTs embedded in the soil (Figure 9.6). The clay sample was contained within an externally insulated strongbox.

9.7   Conclusions

1. Cold regions' engineering encompasses a range of problems; to date, only a few have been investigated using centrifuge models. On the basis of

experimental evidence and a study of the scaling laws, it is possible to make a number of conclusions about the confidence with which centrifuge tests may be undertaken and the benefits that such tests may have over similar small-scale laboratory models. Some cold regions' problems are known to be suitable for centrifuge modelling. For other problems further fundamental investigation is necessary in order to determine the potential benefits, which at the present time can only be suggested.

2. A study of the scaling laws relevant to cold regions' problems indicates a general consistency between processes except with regard to:
   (a) time-dependent processes involving coupled diffusion, inertial effects, strain rate and creep phenomena; and
   (b) the length scaling of the frozen fringe during frost heave, ice crystal size, and crack length in brittle fracture.
   Caution is required where coupled processes take place involving different time or length scales.

3. On examining the processes taking place in cold regions' problems, the following summary can be made regarding the scaling and the benefits of centrifuge modelling.
   (a) Experimental evidence exists for the scaled modelling of sea ice on the centrifuge, and for the scaling of the strength of the resulting ice. The formation of the ice takes advantage of the time compression in the centrifuge, while the deformation testing takes advantage of the correct structure and self-weight of the ice, though the test is likely to have to follow strain rate time scaling, and perhaps fracture mechanics length scaling.
   (b) Thawing events are likely to scale consistently in the centrifuge and take full advantage of the self-weight and time compression benefit of centrifuge modelling. The major exception is where creep of frozen soil within the model may be significant.
   (c) The benefit of centrifuge modelling is also likely to be seen for other problems which involve bodies of both frozen and unfrozen material in which self-weight effects are important, such as pipeline frost heave. In this case, the scaling of frost heave needs further investigation with regard to the effects of creep and the size of the frozen fringe.
   (d) There are several aspects of frozen soil behaviour that may benefit from centrifuge modelling such as the modelling of ultimate strength, and creep processes where self-weight effects are important. However, the time compression benefits of the centrifuge are unlikely to be operative in the latter case.

4. Fundamental studies of cold regions' behaviour are required for: the scaling of creep response, frozen fringe magnitude, ice crystal size in frozen soil, frost heave, crack propagation in frozen material, and the effect of freezing in the laboratory or in flight, on the structure and constitutive response of a frozen soil.

5. Two distinct problem types have been identified: models requiring large temperature variations and phase change with large magnitudes of heat extraction or input, and those requiring fairly steady-state conditions with heat input/output necessary only to overcome insulation losses. In order to correctly simulate the large heat transfer requirements of the former case, it is necessary to develop equipment capable of achieving very high freeze rates since to correctly model a 1/Nth scale model, heat flux rates N times larger than in the prototype must be achieved.

6. In order to achieve freezing and thawing in flight, heat and cold sources that will function on a centrifuge are necessary. Several have been discussed. It is not possible to recommend a single source as each particular one will have its own advantages best suited to a particular modelling requirement.

7. The transfer, at a sufficient rate, of heat or cold to a free model surface is seen as one of the biggest problems in modelling. Direct heat transfer (forced convection) is likely to give the highest flux rates but the poorest uniformity, while indirect heat transfer will give lower flux rates but better uniformity. In either case the achievement of temperatures very much below the final desired model temperature was identified as necessary.

8. The uses of certain sources of cold are dependent on slip-ring capability. Suggested specifications are listed for consideration where it is possible to replace or update slip rings.

9. The centrifuge modelling of cold regions' problems can lead to potential safety hazards not often associated with conventional centrifuge tests. Various safety considerations relating to cold regions modelling have been listed. Considerations relating to package design and ancillary equipment required for model preparation have been discussed and examples of tests reported in the literature are given.

10. In conclusion, the centrifuge modelling of cold regions problems is a promising area: there is scope for many avenues of investigation and a wide range of modelling techniques are possible.

Acknowledgements

The author is grateful to Professor A.N. Schofield for his comments and suggestions made during the preparation of this chapter.

Appendix: Heat transfer to the model surface by conduction, natural convection and radiation

This appendix concerns the indirect heat transfer of energy from a heat exchanging plate to a model surface through a small air gap. It considers conduction, convection and radiation.

Pure conduction between model surface and heat exchanging plate

Let the temperature of the top plate be T_p, the temperature of soil surface be T_s, the gap between the two be H, and the thermal conductivity of the air be k, then the heat flux due to pure conduction is given by:

$$q_{cond} = \frac{k\Delta T}{H}$$

where

$$\Delta T = T_s - T_p$$

For example if $T_p = -50°C$, $T_s = 0°C$ and $H = 0.02\,m$, then for $k = 0.022\,W/mK$,

$$q_{cond} = 55\,W/m^2$$

Natural convection between soil surface and cold plate

If $T_s > T_p$ then natural convection will take place. The process is governed by the Rayleigh number (Holman, 1981, p. 291):

$$Ra_H = \frac{Ng\beta H^3 \Delta T}{\alpha v}$$

where Ng is the centrifuge acceleration, β is the thermal coefficient of expansion of air, α is the thermal diffusivity of air and v is the viscosity of air.
    For an ideal gas, $\beta = 1/T$ where T is in Kelvin. Hence:

$$Ra_H \approx \frac{NgH^3 \Delta T}{\alpha v \bar{T}}$$

where $\bar{T} = (T_s + T_p)/2$.
    An empirical correlation can be given relating the Nusselt number, the ratio of convective heat transfer to conductive heat transfer, to the Rayleigh number as follows:

$$Nu = \frac{q_{conv}}{q_{cond}} = C(Ra_H)^m$$

where for parallel horizontal plates, correlations are as follows:

Ra_H	C	m
<1700	1	0
1700–7000	0.059	0.4
7000–3.2×10^5	0.212	0.25
>3.2×10^5	0.061	0.33

where $0.5 < v/\alpha < 2$, and for the case of $T_p > T_s$, $C = 1$, $m = 0$.

Hence

$$q_{conv} = Nu \cdot q_{cond}$$

For example if $T_p = -50°C$, $T_s = 0°C$, $H = 0.02\,m$, $N = 50$, $\rho = 1.41\,kg/m^3$, $v = 9.49 \times 10^{-6}\,m^2/s$, $k = 0.022\,W/mK$, $\alpha = 0.132 \times 10^{-4}\,m^2/s$ then:

$$Ra_H = 6.32 \times 10^6,$$

$$Nu = 11.3,$$

$$q_{conv} \approx 620\,W/m^2$$

It is seen that for Rayleigh numbers greater than 3.2×10^5, the heat transfer by convection is independent of the height H, permitting a large air gap to be maintained between the soil surface and top plate in these circumstances. It must be noted that in a centrifuge test involving a free water surface, this surface will be curved, and this may need to be accounted for in the shape of the top plate. In addition the results will be modified somewhat by the presence of Coriolis effects.

Radiative heat transfer

The calculation of the magnitude of heat transfer occurring if both surfaces can be considered as black bodies is given by:

$$q_{rad} = \sigma(T_2^4 - T_1^4)F$$

where $\sigma = 56.7 \times 10^{-9}\,W/m^2\,K^4$, and F is the transmittance coefficient which depends on the nature of the transmitting and receiving surfaces $(0 < F < 1)$.

Assuming perfect transmittance $(F = 1)$, taking $T_1 = -50°C = 223\,K$, $T_2 = 0°C = 273\,K$, and assuming a plate separation small relative to the plate area then

$$q_{rad} \approx 175\,W/m^2$$

Combined convective, conductive and radiative heat transfer

Where the model surface is warmer than the heat exchanging plate, then the maximum heat transfer that can be obtained is given by $q_{conv} + q_{rad}$. However in circumstances where the plate is warmer than the model then convection cannot take place and heat transfer is by conduction and radiation only: $q_{cond} + q_{rad}$.

In centrifuge models, the intent is typically to simulate a prototype system. If the prototype heat flux q_{prot}, is known, then the required model heat flux rate is Nq_{prot} where N is the scale factor. Hence it is of value to plot the relationship between q_{prot} and the test acceleration for various values of ΔT so that for a given q_{prot} and test acceleration, the required temperature difference ΔT can be determined. A plot is given in Figure 9.1 for a plate separation of

0.05 m where the soil surface temperature is taken as 0°C. It can be seen that smaller temperature differences are needed for cooling as compared to heating, e.g. a required prototype heat flux of 15 W/m² at a model acceleration of 50 g requires a ΔT of -50°C for cooling and 95°C for heating.

References

Ashby, M.F. and Jones, D.R.H. (1980) Engineering materials: An introduction to their properties and applications. *Int. Series Mater. Sci. Technol.*, **34**, Pergamon Press, Oxford.

Chen, X., Schofield, A.N. and Smith, C.C. (1993) Preliminary tests of heave and settlement of soils undergoing one cycle of freeze–thaw in a closed system on a small centrifuge. *Proc. 6th Int. Conf. Permafrost, Beijing*, Vol. 2, pp. 1070–1072.

Chen, X. and Smith, C.C. (1993) *Frost Heave of Pipelines; Centrifuge and 1 g Model Tests*. Cambridge University Technical Report (CUED/D-SOILS/TR264).

Clough, H.F. and Vinson, T.S. (1986) Centrifuge model experiments to determine ice forces on vertical cylindrical structures. *Cold Reg. Sci. Technol.*, **12**, 245–259.

Clough, H.F., Wurst, P.L. and Vinson, T.S. (1986) Determination of ice forces with centrifuge models. *Geotech. Test. J., ASTM*, **9** (2), 49–60.

Holman, J.P. (1981) *Heat Transfer*, McGraw-Hill, New York.

Jessberger, H.L. (1989) Opening address. *Ground Freezing '88* (eds R.H. Jones and J.T. Holden), Vol. 2, pp. 407–411. Balkema, Rotterdam.

Jessberger, H.L. and Güttler, U. (1988) Bochum geotechnical centrifuge. *Centrifuge '88* (ed. J-F. Corté), pp. 37–44. Balkema, Rotterdam.

Jeyatharan, K. (1992) Dynamic Response of a Caisson Loaded by Ice. PhD Thesis, University of Cambridge.

Ketcham, S.A. (1990) *Applications of Centrifuge Testing to Cold Regions Geotechnical Studies*. USA Cold Regions Research and Engineering Laboratory.

Lach, P.R., Clark, J.I. and Poorooshasb, F. (1993) Centrifuge modelling of ice scour. *Fourth Canad. Conf. Marine Geotechnical Engineering*, pp. 356–374. C-CORE, St. John's, Newfoundland.

Langhorne, P.J. and Robinson, W.H. (1983) Effect of acceleration on sea ice growth. *Nature*, **305**, 695–698.

Lovell, M.S. and Schofield, A.N. (1986) Centrifugal modelling of sea ice. *Proc. First Int. Conf. Ice Technol.*, pp. 105–113.

Malushitsky, Y.N. (1981) *The Centrifugal Model Testing of Waste-heap Embankments*. Cambridge University Press, Cambridge.

Otten, E. (1958) Producing cold air—simplicity of the vortex tube method. *Engineering*, **186**, 154–156.

Palmer, A. (1991) Centrifuge modelling of ice and brittle materials. *Canad. Geotech. J.*, **28**, 896–898.

Palmer, A.C., Schofield, A.N., Vinson, T.S. and Wadhams, P. (1985) Centrifuge modelling of underwater permafrost and sea ice. *Proc. 4th Int. Offshore Mechanics and Arctic Engineering Symposium, Texas*, Vol. II, pp. 65–69.

Phillips, R. (1985) *Data Report of Centrifuge Model Tests XS1-XS6*. Andrew N. Schofield & Associates. Report to R.J. Brown & Associates, November 1985.

Schofield, A.N. (1980) Cambridge geotechnical centrifuge operations; Twentieth Rankine Lecture, *Géotechnique*, **30** (3), 227–268.

Schofield, A.N. and Taylor, R.N. (1988) Development of standard geotechnical centrifuge operations. *Centrifuge '88* (ed. J.-F. Corté), pp. 29–32. Balkema, Rotterdam.

Smith, C.C. (1992) Thaw Induced Settlement of Pipelines in Centrifuge Model Tests. PhD Thesis, University of Cambridge.

Vinson, T.S. and Palmer, A.C. (1988) Physical model study of arctic pipeline settlements. *Proc. 5th Int. Conf. Permafrost, Trondheim*, Vol. 2, pp. 1324–1329.

Vinson, T.S. and Wurst, P.L. (1985) Centrifugal modelling of ice forces on single piles. *Civil Engineering in the Arctic Offshore* (eds F.L. Bennett and J.L. Machemehl), ASCE, pp. 489–497.

Index